SWITCHING POWER SUPPLY DESIGN

A CONCISE PRACTICAL HANDBOOK

Lazar Rozenblat

Copyright © 2021 Lazar Rozenblat

All rights reserved.

No part of this book may be copied or reproduced in any format without permission in writing from the copyright holder.
All trademarks belong to their respective owners.
Cover design by pro_ebookcovers

ISBN: 9798757654942

CONTENTS

PREFACE 6

1 SMPS TOPOLOGIES 8

    1.1 CALCULATION OF DC TRANSFER FUNCTION 8

    1.2 BUCK CONVERTER 10

    1.3 FLY-BUCK™ CONVERTER 13

    1.4 BOOST CONVERTER 16

    1.5 BUCK-BOOST (INVERTING) 19

    1.6 SEPIC CONVERTER 22

    1.7 SINGLE-SWITCH FORWARD CONVERTER 25

    1.8 FORWARD CONVERTER VARIATIONS 29

    1.9 FLYBACK CONVERTER IN DCM 31

    1.10 FLYBACK CONVERTER IN CCM 35

    1.11 FLYBACK CONVERTER VARIATIONS 40

    1.12 HALF BRIDGE 42

    1.13 LLC HALF BRIDGE 46

    1.14 PHASE SHIFTED FULL BRIDGE WITH CURRENT DOUBLER 52

    1.15 TOPOLOGY SELECTION 56

2 POWER FACTOR CORRECTION 58

    2.1 POWER FACTOR BASICS 58

2.2   CCM PFC BOOST 60

2.3   DCM PFC BOOST 65

3   FEEDBACK LOOP FUNDAMENTALS 69

3.1   BODE PLOT AND STABILITY CRITERIA 69

3.2   LAPLACE TRANSFORM: ZEROES AND POLES 72

3.3   TRANSIENT RESPONSE vs. PHASE MARGIN 75

3.4   CHOOSING CROSSOVER FREQUENCY 77

3.5   LOOP COMPENSATION BASICS 79

3.6   VOLTAGE MODE CONTROL TRANSFER FUNCTIONS 81

3.7   CURRENT MODE CONTROL TRANSFER FUNCTIONS 83

3.8   INCREASING PHASE MARGIN 87

3.9   ERROR AMPLIFIER GAIN WITH TL431 88

4   MAGNETICS DESIGN 90

4.1   TRANSFORMER CALCULATIONS 90

4.2   CORE LOSSES 93

4.3   COPPER LOSSES 94

4.4   POWER INDUCTOR DESIGN 97

5   MISCELLANEOUS POWER ELECTRONICS REFERENCE 100

5.1   ESTIMATION OF MOSFET LOSSES 100

5.2　CALCULATION OF OUTPUT VOLTAGE RIPPLE 102

5.3　CAPACITANCE CALCULATION FOR LOAD TRANSIENT RESPONSE 103

5.4　PCB TRACE PROPERTIES... 105

6　REFERENCES... 107

7　APPENDIX 109

7.1　MAGNETIC UNIT CONVERSION............................. 109

8　ABOUT THE AUTHOR 110

PREFACE

The intention of this handbook is to provide in a single place all the essential information needed in the practical switching mode power supply (SMPS) design in an easy-to-use format. I hope it will be as useful to the experienced designer as it will to the recent engineering grad, a student, and a hobbyist.

Here are the key covered topics:
- Main practically used isolated and non-isolated converter topologies, including active PFC;
- Power transformer and inductor design and estimation of the losses;
- Feedback control loop relationships including transfer function with TL431;
- Miscellaneous design and analysis topics, such as MOSFET switching time and losses, capacitance calculation for transient response, PCB trace characteristics, and little-known empirical equations.

The covered converter topologies are:
- Buck
- Fly-Buck™
- Boost
- Buck-boost (non-isolated flyback)
- SEPIC
- CCM and DCM isolated flyback
- Forward (including active clamp forward)
- Half-bridge
- Phase shifted full bridge with current doubler
- LLC
- CCM and DCM PFC boost

For each covered topology, I provided power plant diagram, brief operation principal, basic waveforms, DC transfer function with efficiency factor, voltage and current stresses in switches and rectifiers, magnetics equations, DC and AC components of the currents in all coils, and often overlooked RMS currents in input and output capacitors. The analysis is provided for worth case input voltage – something that is not always obvious. For example, for a CCM flyback the transformer's magnetizing inductance has to be calculated at high line, while for DCM – at low line.

Note that this is not a textbook for learning power electronics. This handbook is for those who know the electronics basics and need a quick reference and practical engineering equations. It should speed up your design by saving time that would otherwise be spent on deriving equations and searching the literature, not to mention on re-spinning the board because of incorrectly selected magnetics, underrated components, or improperly sized PCB traces.

The formulas are presented with brief explanations but without detailed derivations.

The magnetic equations are given in CGS system. The Appendix provides the conversion between SI and CGS. Knowing how frustrating it can be to search through a book for definitions and units of quantities used in equations, definitions and units of all quantities are repeated in each chapter.

I make no representation that the use of any topology described in this book will not infringe on existing or future patent rights or other rights, and I am not implying that I grant license to use them.

1 SMPS TOPOLOGIES

1.1 CALCULATION OF DC TRANSFER FUNCTION

DC-DC converter's output voltage is calculated based on the fact that in steady state the net volt-seconds across any inductor and net amp-seconds of any capacitor over one switching cycle must be zero.

Example. DC transfer function calculations for buck converter operating in continuous conduction mode at constant frequency:

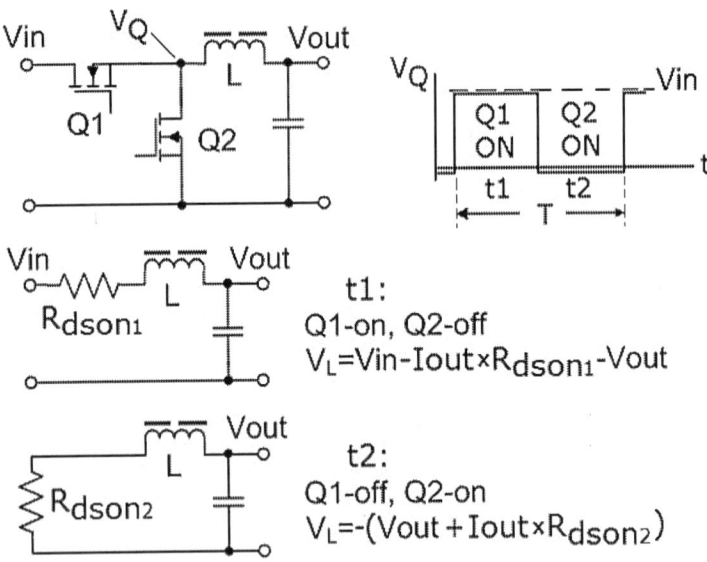

The switching cycle T has two sub-intervals t1 and t2. During t1 switch Q1 is ON, Q2 is OFF; during t2: Q1 is OFF and Q2 is ON.

Neglecting voltage and current ripple, volt-second balance is:

$$(V_{in} - I_{OUT} \cdot R_{dson1} - V_{OUT}) \cdot t1 = (V_{OUT} + I_{OUT} \cdot R_{dson2}) \cdot t2$$

where R_{dson1} and R_{dson2} are the drain to source resistances of Q1 and Q2 during their "on" state.
Solving for Vout:

$$V_{out} = V_{in} \cdot D - I_{OUT} \cdot (D \cdot R_{dson1} - R_{dson2} + D \cdot R_{dson2}),$$

where $D \overset{def}{=} t1/T$ – duty cycle.

The DC transfer function **Vout/Vin** generally is a function of the load due to power losses. However, in most textbooks power losses and accordingly the second term in the above equation are neglected, and output is expressed as **Vout=Vin·D**. Such an equation is easy to memorize and to deal with, but it results in up to 10% error since it ignores the fact that the converter has to process the dissipated energy. A more practical simplified equation taking into account power losses is:

$$V_{out} \approx \eta \cdot V_{in} \cdot D$$

where η – expected efficiency of the converter (typically 0.85 to 0.96).

All the equations in this book will similarly include the expected efficiency factor, which indirectly includes the effect of voltage drops in switches and rectifiers. However, for low voltage outputs you may still need to add actual rectifier drop to **Vout**.

1.2 BUCK CONVERTER

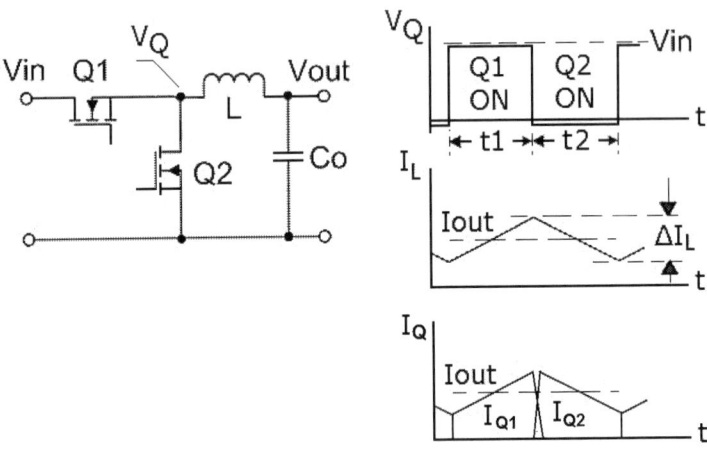

When switch Q1 is ON, energy from input source is transferred to the output via inductor L. In the process, some energy is also accumulated in the inductor. When Q1 is OFF and Q2 is ON, inductor continues conducting via Q2 and releases all or a portion of its stored energy into the load and to output capacitor Co. Lower-cost lower-efficiency designs have a diode instead of Q2. Output voltage is always lower than input voltage: **Vout<Vin**.

OUTPUT VOLTAGE

In continuous conduction mode (CCM): **Vout=η·Vin·D**, where η - efficiency of the converter, D=t1/T - duty cycle of the switch Q1 (D<1); T=t1+t2 - switching period.

INDUCTOR L

Average current I_{L_AV}=Iout;
Peak-to-peak ripple current: ΔI$_L$=Vout·(1-D)/(L·F), where F=1/T - frequency in hertz.
RMS value of inductor ripple current: $I_{L\_AC\_RMS}$=ΔI$_L$/√12.
Maximum current ripple appears at high line:

$$\Delta I_{L\_MAX} = \frac{Vout \cdot (1 - \frac{Vout}{\eta \cdot V_{IN\_MAX}})}{L \cdot F}$$

where V_{IN_MAX} - maximum input voltage.

The condition of CCM is $\Delta I_L/2 < I_{out}$. If CCM is desired over entire input voltage range, the required inductance:

$$L_{MIN} > V_{out} \cdot (1 - V_{out}/\eta V_{IN\_MAX})/(2F \cdot I_{out})$$

In practice, L is selected several times larger than L_{MIN}, so that $\Delta I_{L\_MAX} \leq (0.3...0.6) \cdot I_{L\_AV}$.

SWITCH Q1
Peak current: $I_{Q1\_PK} = I_{out} + \Delta I_L/2$;
Center value of current pulse (midpoint of the ramp): $I_{CV1} = I_{out}$;
RMS current (neglecting ΔI_L): $I_{Q1\_RMS} = I_{out} \cdot \sqrt{(V_{out}/\eta V_{IN})}$;
Peak voltage: $V_{Q1\_PK} = V_{in}$;
Maximum conduction losses appear at low line. Neglecting ΔI_L:
$$P_{Q1\_COND} \approx I_{out}^2 \cdot R_{dson1} \cdot V_{out}/(\eta V_{IN\_MIN}),$$
where R_{dson1} - Q1 drain to source resistance in on-state.

If PWM controller limits Q1 peak current to I_{PK_LIM}, maximum achievable load current: $I_{OUT\_MAX} = I_{PK\_LIM} - \Delta I_L/2$.

SWITCH Q2
Peak current: $I_{Q2\_PK} = I_{out} + \Delta I_L/2$;
Center value of current pulse: $I_{CV2} = I_{out}$;
RMS current (neglecting ΔI_L): $I_{Q2\_RMS} = I_{out} \cdot \sqrt{(1 - V_{out}/\eta V_{IN})}$;
Peak voltage: $V_{Q2\_PK} = V_{in}$;
Maximum conduction losses appear at high line:
$$P_{Q2\_COND} \approx I_{out}^2 \cdot R_{dson2} \cdot (1 - V_{out}/\eta V_{IN\_MAX}),$$
where R_{dson2} - Q2 drain to source resistance in on-state.

OUTPUT CAPACITOR Co
Peak to peak current: $\Delta I_C = \Delta I_L$;
RMS current: $I_{C\_rms} = I_{L\_AC\_rms}$.
Net output capacitance is selected to satisfy ripple requirements (see Chapter 5.2) and transient response (see Chapter 5.3).

INPUT CURRENT

Average input current: $I_{IN\_AV} = I_{out} \cdot V_{out}/\eta V_{IN}$.

RMS value of AC component of input current, which is current through input capacitor (not shown on the diagram):

$$I_{IN\_RMS} = \sqrt{\frac{V_{out}}{\eta V_{in}} \cdot \left[I_{out}^2 \cdot \left(1 - \frac{V_{out}}{\eta V_{in}}\right) + \frac{\Delta I_L^2}{12} \right]}$$

If $\Delta I_L \ll I_{out}$:

$$I_{IN\_AC\_RMS} \approx I_{out} \cdot \sqrt{\frac{V_{out}}{\eta V_{in}} \cdot \left(1 - \frac{V_{out}}{\eta V_{in}}\right)}$$

The AC component of input current reaches maximum $I_{IN\_AC\_RMS\_MAX} \approx 0.5 \cdot I_{out}$ when $V_{in} = 2 \cdot V_{out}/\eta$.

1.3 FLY-BUCK™ CONVERTER

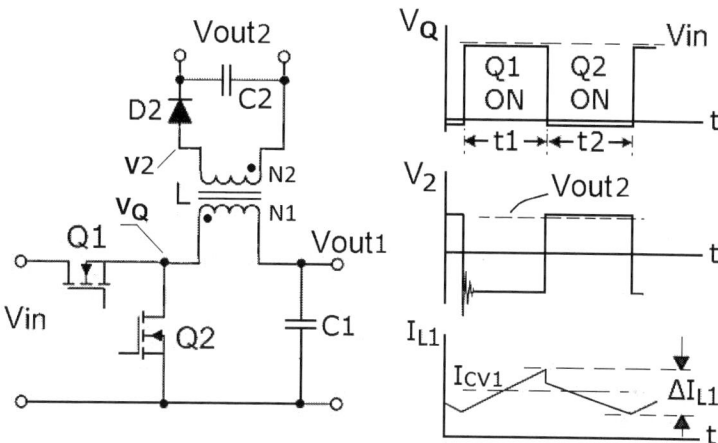

Fly-Buck™ is a trademark of Texas Instrument Inc.

This is a buck converter with auxiliary inductor's winding that provides second flyback-like output. When switch Q1 is ON, switch Q2 and diode D2 are OFF, and energy from input source is transferred to the output 1 via inductor L. In the process, some energy is also accumulated in its core. When Q1 is OFF, inductor continues conducting current I_{L1} into output 1 via Q2 and releases a portion of its stored energy into output 2 via its secondary coil and diode D2. Lower-cost lower-efficiency designs have a diode instead of Q2. Output voltage **Vout1** is always lower than input voltage: Vout1<Vin; auxiliary voltage **Vout2** can be both lower and higher than Vin depending on the inductor turns ratio.

OUTPUT VOLTAGE

In continuous conduction mode (CCM): **Vout1=η·Vin·D**, where η - efficiency of the converter, D=t1/T - duty cycle of the switch Q1 (D<1); T=t1+t2=1/F – switching period, F- switching frequency is hertz.

Auxiliary voltage (neglecting voltage drop in Q2):
Vout2 ≈ (Vout1·N2/N1) - Vrect, where N1 and N2 – inductor's primary and secondary turns, **Vrect** – voltage drop of diode D2.

INDUCTOR L

Average current in primary coil **N1** (center value of the ramp):
$$I_{CV1} = I_{out1} + I_{out2} \cdot V_{out2}/(\eta_2 \cdot V_{out1}),$$
where **Iout1** and **Iout2** – load currents on outputs 1 and 2 respectively; η_2 – efficiency of **Vout2** channel.

Peak-to-peak ripple current: $\Delta I_{L1} = V_{out1} \cdot (1-D)/(L \cdot F)$;
Maximum current ripple appears at high line:
$$\Delta I_{L1\_MAX} = \frac{V_{out1} \cdot \left(1 - \frac{V_{out1}}{\eta \cdot V_{IN\_MAX}}\right)}{L \cdot F}$$
where V_{IN_MAX} - maximum input voltage.
RMS value of primary coil's AC ripple current: $I_{L1\_RMS} \approx \Delta I_{L1}/\sqrt{12}$.

The condition of CCM is $\Delta I_{L1}/2 < I_{CV1}$. If CCM is desired over entire input voltage range, the required primary inductance:

$$L_{MIN} > V_{out1} \cdot (1 - V_{out1}/\eta V_{IN\_MAX})/(2F \cdot I_{CV1})$$

In practice, L is selected 3 to 6 times larger than L_{MIN}, so that $\Delta I_{L1\_MAX} \leq (0.3...0.6) \cdot I_{L\_CV1}$.

Required turns ratio: N1/N2 ≈ Vout1/(Vout2+Vrect).

SWITCH Q1

Peak current: $I_{Q1\_PK} = I_{CV1} + \Delta I_{L1}/2$;
RMS current (neglecting ΔI_{L1}): $I_{Q1\_RMS} = I_{CV1} \cdot \sqrt{(V_{out1}/\eta V_{IN})}$;
Peak voltage: $V_{Q1\_PK} = V_{in}$;
Maximum conduction losses appear at low line. Neglecting ΔI_{L1}:
$$P_{Q1\_COND} \approx I_{CV}^2 \cdot R_{dson1} \cdot V_{out1}/(\eta V_{IN\_MIN}),$$
where R_{dson1} - Q1 drain to source resistance in on-state.

SWITCH Q2

Peak current: $I_{Q2\_PK} = I_{out1} + \Delta I_{L1}/2$;

Center value of current pulse: $I_{cv2} = Iout1$;
RMS current (neglecting ΔI_{L1}): $I_{Q2\_RMS} = Iout1 \cdot \sqrt{(1-Vout1/\eta V_{IN})}$;
Peak voltage: $V_{Q2\_PK} = Vin$;
Maximum conduction losses appear at high line:
$$P_{Q2\_COND} \approx Iout1^2 \cdot R_{dson2} \cdot (1-Vout1/\eta V_{IN\_MAX}),$$
where R_{dson2} - Q2 drain to source resistance in on-state.

OUTPUT CAPACITOR Co

Peak to peak current: $\Delta I_C = \Delta I_{L1}$;
RMS current: $Ic\_rms = I_{L1}\_rms$.
Net output capacitance is selected to satisfy ripple requirements (see Chapter 5.2) and transient response (see Chapter 5.3).

DIODE D1

Peak voltage (plateau): $V_{D1\_PK} = Vout2 + Vout1 \cdot N2/N1$
Average current: $I_{D1\_AV} = Iout2$

INPUT CURRENT

Average input current:
$I_{IN\_AV} = (Iout1 \cdot Vout1 + Iout2 \cdot Vout2)/\eta V_{IN}$.
RMS value of AC component of input current, which is current through input capacitor (not shown on the diagram):

$$I_{IN\_AC\_RMS} \approx \sqrt{\frac{Vout1}{\eta Vin} \cdot \left[Icv_1^2 \cdot \left(1 - \frac{Vout1}{\eta Vin}\right) + \frac{\Delta I_{L1}^2}{12} \right]}$$

If $\Delta I_{L1} \ll Icv_1$:

$$I_{IN\_AC\_RMS} \approx Icv_1 \cdot \sqrt{\frac{Vout1}{\eta Vin} \cdot \left(1 - \frac{Vout1}{\eta Vin}\right)}$$

The AC component of input current reaches maximum when $Vin = 2 \cdot Vout1/\eta$: $I_{IN\_AC\_RMS\_MAX} \approx 0.5 \cdot Icv_1$

1.4 BOOST CONVERTER

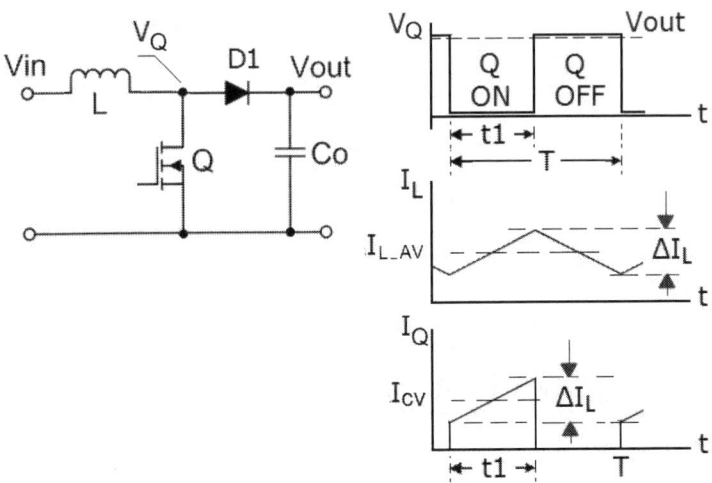

When switch Q is ON, D1 is OFF, and energy from input source is accumulated in the inductor L. When Q is OFF, inductor is continuing conducting current through diode D1 and releases all or a portion of its stored energy to the load and to output capacitor Co. Output voltage is always higher than input: **Vout>Vin**. In continuous conduction mode (CCM) the inductor current I_L does not reach zero.

OUTPUT VOLTAGE
In CCM:
$$Vout = \eta Vin/(1 - D)$$
where η- converter efficiency; D=t1/T - duty cycle (D<1).

INDUCTOR L
Average current I_{L_AV}=Vout·Iout/(ηVin);
Peak to peak ripple current:
$$\Delta I_L = Vin \cdot (1-\eta Vin/Vout)/(L \cdot F),$$
where F- frequency is hertz, L – inductance in henry.

RMS value of inductor ripple current: $I_{L\_RMS} = \Delta I_L/\sqrt{12}$.

Maximum current ripple $\Delta I_{L\_MAX}=Vo/(4\eta \cdot L \cdot F)$ appears when $Vin=Vout/2\eta$.
The condition of CCM: $\Delta I_L/2 < I_{L\_AV}$. This requires:
$$L > \eta \cdot Vin^2 \cdot D/(2Po \cdot F)$$
where $Po=Vout \cdot Iout$ – output power.
If CCM is desired over entire input voltage range:

$$L_{MIN} > \frac{\eta V_{IN\_MAX}^2 \cdot (1 - \eta V_{IN\_MAX}/Vout)}{2F \cdot Po}$$

where V_{IN_MAX} - maximum input voltage.
In practice, L is selected 3 to 6 times larger than L_{MIN}, so that $\Delta I_{L\_MAX} \le (0.3...0.6) \cdot I_{L\_AV}$.

SWITCH Q

Center value of current pulse (midpoint of the ramp):
$Icv = I_{L\_AV} = Iout \cdot Vout/\eta Vin$;
Peak current: $I_{Q\_PK}=Iout \cdot Vout/\eta Vin + \Delta I_L/2$.
RMS current:

$$I_{Q\_RMS} = \sqrt{D \cdot (I_{CV}^2 + \frac{\Delta I_L^2}{12})}$$

If $\Delta I_L < Icv$:

$$I_{Q\_RMS} \approx \frac{Iout \cdot Vout}{\eta Vin} \cdot \sqrt{1 - \frac{\eta Vin}{Vout}}$$

Peak voltage: $V_{Q\_PK}=Vout$;
Conduction losses are highest at low line:

$$P_{Q\_COND} \approx \left(\frac{Iout \cdot Vout}{\eta V_{IN\_MIN}}\right)^2 \cdot (1 - \frac{\eta V_{IN\_MIN}}{Vout}) \cdot R_{DS\_ON}$$

where R_{dson} - drain to source resistance of Q in on-state.

If PWM controller limits switch peak current to I_{PK_LIM}, maximum achievable load current: $I_{OUT\_MAX}=(I_{PK\_LIM} - \Delta I_L/2) \cdot \eta Vin/Vout$.

RECTIFIER D
Peak current: $I_{D\_PK} = I_{out} \cdot V_{out}/\eta V_{in} + \Delta I_L/2$;
Average current: $I_{D\_AV} = I_{out}$;
Peak reverse voltage: $V_{D\_PK} = V_{out}$;
Conduction losses: $P_{D\_COND} \approx I_{out} \cdot V_D$,
where V_D- diode's forward voltage drop, typically 0.5-1.0V.

OUTPUT CAPACITOR Co
Instantaneous current:

$I_C = -I_{out}$ during t1;
$I_C = I_L - I_{out}$ between t1 and T,
where $T = 1/F$ – switching period.
RMS value of current (neglecting ΔI_L):

$$I_{C\_RMS} = I_{out} \cdot \sqrt{\frac{\eta V_{in}}{V_{out}} - \left(\frac{V_{out}}{\eta V_{in}} - 1\right)^2 \cdot \left(\frac{\eta V_{in}}{V_{out}} - 1\right)}$$

Net output capacitance is selected to satisfy ripple requirements (see Chapter 5.2) and transient response (see Chapter 5.3).

INPUT CURRENT
Average input current: $I_{IN\_AV} = I_{out} \cdot V_{out}/\eta V_{in}$.
RMS value of AC component of input current, which is current through input capacitor (not shown on the diagram):
$I_{IN\_AC\_RMS} = I_{L\_RMS}$.
It reaches maximum value at $V_{in} = V_{out}/2\eta$:
$I_{IN\_AC\_RMS\_MAX} = V_o/(8\eta \cdot L \cdot F \cdot \sqrt{3})$.

1.5 BUCK-BOOST (INVERTING)

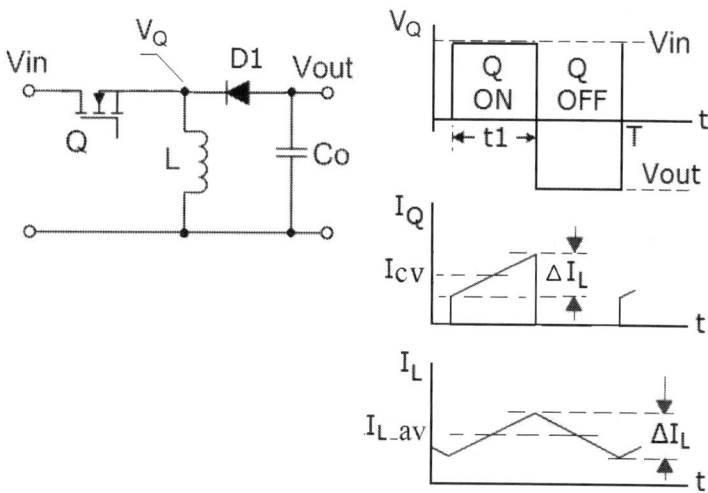

During ON state of the switch Q energy is accumulated in the inductor. During OFF state the inductor voltage reverses polarity and releases all or a portion of its stored energy via diode D1 into the load and output capacitor Co. Output voltage is negative and its absolute value can be both below and above input voltage depending on the duty cycle. In continuous conduction mode (CCM) the inductor current I_L does not reach zero.

OUTPUT VOLTAGE
In CCM:
$$Vout = \eta Vin \cdot \frac{D}{1-D}$$
where **Vout** – absolute value of output voltage (which is negative), η - converter efficiency; $D = t1/T$ - duty cycle ($D<1$).
In steady state operation:
$$D = \frac{Vout}{\eta Vin + Vout}$$

INDUCTOR L
Average current $I_{L\_AV} = Iout/(1-D)$;
Maximum peak to peak ripple current appears at high line:

$$\Delta I_{L\_MAX} = \frac{Vin \cdot D}{L \cdot F} = \frac{Vout}{L \cdot F} \cdot \frac{1}{\frac{Vout}{V_{IN\_MAX}} + \eta}$$

where F- frequency is hertz, L – inductance in henry.
RMS value of inductor ripple current: $I_{L\_RMS} = \Delta I_L/\sqrt{12}$
The condition of CCM: $\Delta I_L/2 < I_{L\_AV}$, which requires

$$L > \frac{\eta \cdot Vin^2 \cdot D^2}{2 \cdot Pout \cdot F}$$

where $Po = Vout \cdot Iout$ – output power.
If CCM is desired over entire input voltage range:

$$L_{MIN} > \frac{\eta \cdot V_{IN\_MAX}^2 \cdot D^2}{2 \cdot Pout \cdot F} = \frac{\eta V_{IN\_MAX}^2 \cdot Vout}{2F \cdot Iout \cdot (Vout + \eta V_{IN\_MAX})^2}$$

where V_{IN_MAX} - maximum input voltage.
In practice, L is selected several times larger than L_{MIN}, such that $\Delta I_L \leq (0.3...0.6) \cdot I_{L\_AV}$.

SWITCH Q

From power balance, center value of current pulse (midpoint of the ramp):

$$Icv = \frac{Iout \cdot Vout}{\eta V_{IN} \cdot D} = Iout \cdot (\frac{Vout}{\eta V_{IN}} + 1)$$

Peak current: $I_{Q\_PK} = Icv + \Delta I_L/2$;
Peak voltage: $V_{Q\_PK} = Vin + Vout$;
RMS current (neglecting ΔI_L): $I_{Q\_RMS} \approx Icv \cdot \sqrt{D}$.
Conduction losses: $P_{Q\_COND} \approx Icv^2 \cdot R_{dson} \cdot D$,

where R_{dson} - drain to source resistance of Q in on-state.
Conduction losses reach maximum at low line:

$$P_{Q\_COND} \approx Iout^2 \cdot R_{DSON} \cdot \frac{Vout \cdot (\eta V_{IN\_MIN} + Vout)}{(\eta V_{IN\_MIN})^2}$$

where V_{IN_MIN} - minimum input voltage.

If PWM controller limits switch peak current to I_{PK_LIM}, maximum achievable load current: $I_{OUT\_MAX}=(I_{PK\_LIM}-\Delta I_L/2)/(1+V_{out}/\eta V_{in})$.

RECTIFIER D1

Center value of current pulse (midpoint of the ramp): $I_{D\_CV}=I_{out}$;
Current averaged over entire period: $I_{D\_AV}=I_{out}\cdot(1-D)$;
Peak current: $I_{D\_PK}=I_{out}+\Delta I_L/2$;
Peak reverse voltage: $V_{D\_PK}=V_{out}+V_{in}$;
Conduction losses: $P_{D\_COND}\approx I_{out}\cdot V_D\cdot(1-D)$,
where V_D- diode's forward voltage drop, typically 0.5-1.0V.

OUTPUT CAPACITOR Co

Current:

$I_C=-I_{out}$ during t1
$I_C=I_L-I_{out}$ (between t1 and T),
where $T=1/F$ – switching period.

Maximum RMS value of current (neglecting ΔI_L):

$$I_{Co\_RMS} = I_{out}\cdot\sqrt{\frac{V_{out}}{\eta V_{IN\_MIN}}}$$

Net output capacitance is selected to satisfy ripple requirements (see Chapter 5.2) and transient response (see Chapter 5.3).

INPUT CURRENT

Average input current: $I_{IN\_AV}=I_{out}\cdot V_{out}/\eta V_{in}$.
Neglecting ΔI_L, maximum value of AC component of input current, which is current through input capacitor (not shown on the diagram):

$$I_{IN\_AC\_RMS} = I_{out}\cdot\sqrt{\frac{V_{out}}{\eta V_{IN\_MIN}}}$$

1.6 SEPIC CONVERTER

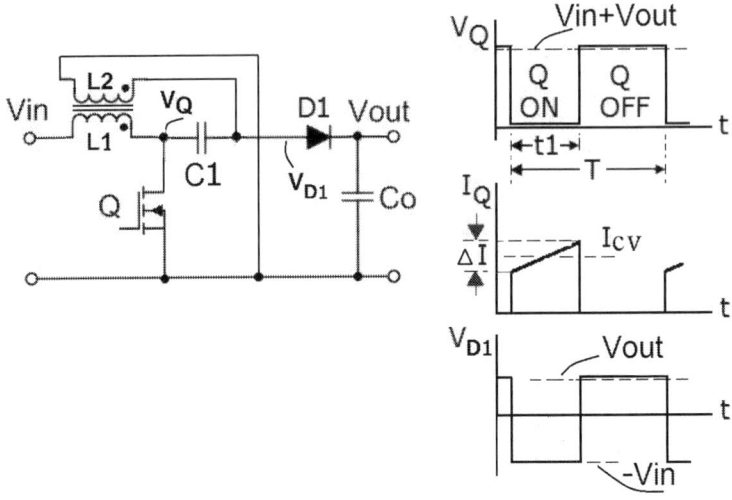

When switch Q is ON, D1 is OFF, and energy from input source is accumulated in the inductor L1. At the same time, C1 (which is charged to Vin) is connected in parallel with L2. When Q1 is OFF, L1 current continues to flow through C1 and D1 into the load and Co. Output voltage can be both below and above input depending on duty cycle. The following equations are given for a coupled inductor.

OUTPUT VOLTAGE
In continuous conduction mode (CCM):

$$\text{Vout} = \eta \cdot \text{Vin} \cdot \frac{D}{1-D}$$

where η- converter efficiency; $D=t1/T$ - duty cycle of the switch Q:

$$D = \text{Vout}/(\text{Vout} + \eta\text{Vin})$$

INDUCTORS L1, L2
Average current:
$I_{L1\_AV} = \text{Vout} \cdot \text{Iout}/\eta\text{Vin}$;
$I_{L2\_AV} = \text{Iout}$

In an ideal coupled inductor with equal turns L1=L2. Mutual inductance causes the ripple current to be split between the two inductors, so that effective inductance L=2·L1=2·L2. Peak to peak ripple current in each inductor:

$$\Delta I_L = \frac{Vin}{L \cdot F} \cdot \frac{Vout}{Vout + \eta \cdot Vin}$$

where L – inductance in henry.
Note: with separate inductors L=L1=L2, so ΔI_L is twice larger.
The condition of continuous conduction mode (CCM): $\Delta I_L/2 < I_{L\_AV}$.
For CCM over entire input voltage range this requires:

$L_{MIN} > V_{IN\_MAX} \cdot Vout/2F \cdot (Vout + \eta\ V_{IN\_MAX}) \cdot I_{L1\_av}$ if Vin>Vout,

$L_{MIN} > V_{IN\_MAX} \cdot Vout/2F \cdot (Vout + \eta\ V_{IN\_MAX}) \cdot Iout$ if Vin<Vout

where V_{IN_MAX} - maximum input voltage.
In practice, L is usually selected several times larger than L_{MIN}, so that $\Delta I_L \leq (0.3...0.6) \cdot I_{L\_AV}$.
RMS value of inductor ripple current: $I_{L\_RMS} = \Delta I_L/\sqrt{12}$.

SWITCH Q

Peak current: $I_{Q\_PK} = Iout(1+Vout/\eta Vin) + \Delta I_L/2$;
Center value of pulse current (midpoint of the ramp):
$Icv = Iout(1+Vout/\eta Vin)$;
Peak voltage: $V_{Q\_PK} = Vin+Vout$;
RMS current (neglecting ΔI_L):

$$I_{Q\_RMS} \approx \frac{Iout}{\eta Vin} \cdot \sqrt{Vout \cdot (Vout + \eta Vin)}$$

Conduction losses: $P_{Q\_COND} = I_{Q\_RMS}^2 \cdot R_{dson}$, where R_{dson} is drain to source resistance of Q in on-state.

If PWM controller limits switch peak current to I_{PK_LIM}, maximum achievable load current for given Vout:
$I_{OUT\_MAX} = (I_{PK\_LIM} - \Delta I_L/2)/(1 + Vout/\eta Vin)$.

RECTIFIER D1
Peak current: $I_{D\_PK} = I_{out}(1+V_{out}/\eta V_{in}) + \Delta I_L/2$;
Average current: $I_{D\_AV} = I_{out}$;
Peak reverse voltage: $V_{D\_PK} = V_{in} + V_{out}$;
Conduction losses: $P_{D\_COND} \approx I_{out} \cdot V_D$,
where V_D- diode forward voltage drop, typically 0.5-1.0V.

OUTPUT CAPACITOR Co
RMS current (neglecting ΔI_L):
$$I_{Co\_RMS} = I_{out} \cdot \sqrt{V_{out}/V_{in}}$$
Net output capacitance is selected to satisfy ripple requirements (see Chapter 5.2) and transient response (see Chapter 5.3).

COUPLING CAPACITOR C1

RMS current (neglecting ΔI_L):
$$I_{C1\_RMS} = I_{out} \cdot \sqrt{V_{in}/V_{out}}$$

Maximum voltage: $V_{C1\_MAX} = V_{in}$.

INPUT CURRENT
Average input current: $I_{IN\_AV} = V_{out} \cdot I_{out}/\eta V_{in}$.
RMS value of AC component of input current, which is current through input capacitor (not shown on the diagram):
$I_{IN\_AC\_RMS} = I_{L\_RMS}$.

1.7 SINGLE-SWITCH FORWARD CONVERTER

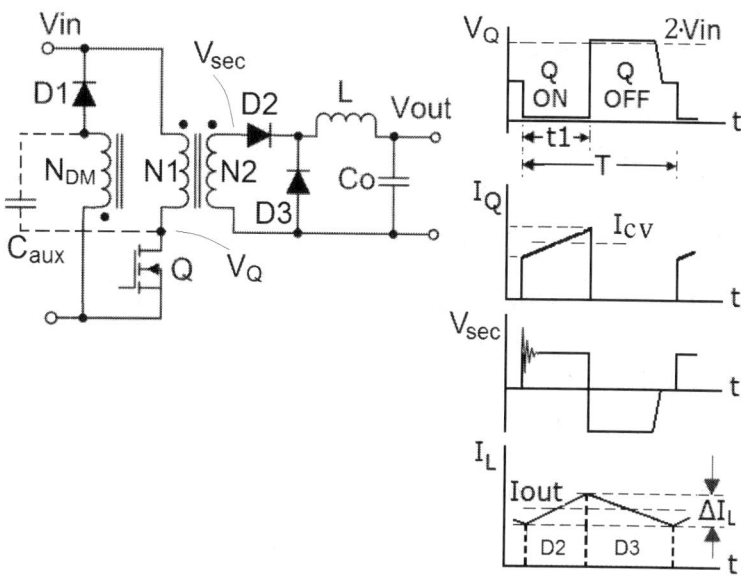

When switch Q is ON, rectifier D2 conducts, and energy from input source is transferred to the output via inductor L. In the process, some energy is also accumulated in the inductor and in transformer's core. When Q is OFF, inductor continues conducting via rectifier D3 and releases a portion of its stored energy into the load. During this time magnetizing current is flowing via D1 and demagnetizing winding (DM), which allows the transformer's core to reset. An auxiliary (optional) capacitor **Caux** clamps voltage spikes caused by leakage inductance.

OUTPUT VOLTAGE
When output inductor operates in continuous conduction mode:

$$Vout = \eta Vin \cdot D \cdot N2/N1$$

where $D=t1/T$- duty cycle of the switch Q ($D<0.5$);
η - efficiency of the converter;
N1 and N2- transformer's primary and secondary turns respectively.

TRANSFORMER

Turns ratio must satisfy the inequation:

$$\frac{N2}{N1} > (V_{out} + V_{rect})/(\eta V_{IN\_MIN} \cdot D_{max})$$

where **Dmax** - maximum duty cycle of PWM; V_{IN_MIN} - minimum input voltage, **Vrect** – rectifier voltage drop.

Minimum required **N1·Ac** product, where **Ac**- core cross-sectional area (in sq.cm):

$$N1 \cdot Ac \geq \frac{V_{IN\_MIN} \cdot D_{max}}{F \cdot (B_{pk} - B_r) \cdot 10^{-8}}$$

where **Bpk** - peak working magnetic flux density in gauss (usually selected 2000-3000 gauss); **Br** - remanence flux density (typically 800-1500 G for un-gapped soft ferrites), **F** – frequency (Hz).

Note. With such **N1·Ac** selection, magnetic flux can exceed **Bpk** during transients. The circuit should employ pulse-by-pulse current limit. For a conservative design, in the numerator of above equation for **N1·Ac** put V_{IN_MAX} instead of V_{IN_MIN}.

Peak magnetizing current: $I\mu = V_{in} \cdot D / F \cdot L\mu$, where $L\mu$ - transformer's magnetizing (primary) inductance in henry.

RMS value of primary current (neglecting ramp of the current pulse):

$$I_{1\_RMS} \approx I_{out} \cdot N2/N1 \cdot \sqrt{D} = I_{out} \cdot \sqrt{\frac{V_o \cdot N2}{\eta V_{in} \cdot N1}}$$

DC component of primary current: $I_{1\_DC} = V_{out} \cdot I_{out}/(\eta \cdot V_{in})$.

RMS value of AC component of primary current:

$$I_{1\_AC\_RMS} = \sqrt{I_{1\_RMS}^2 - I_{1\_DC}^2}$$

Neglecting ramp of the current pulse:

$$I_{1\_AC\_RMS} \approx I_{out} \cdot \sqrt{\frac{V_o}{\eta V_{in}} \cdot \frac{N2}{N1} - \frac{V_o^2}{(\eta V_{in})^2}}$$

If $N2/N1 = V_{out}/(\eta V_{IN\_MIN} \cdot D_{MAX})$, maximum AC component:

$$I_{1\_AC\_RMS} \approx \frac{I_{out} \cdot V_{out}}{\eta V_{IN\_MIN}} \sqrt{\frac{1}{D_{max}} - 1}$$

Maximum RMS value of secondary current:

$$I_{2\_RMS} \approx I_{out} \cdot \sqrt{D} = I_{out} \cdot \sqrt{\frac{V_{out} \cdot N1}{\eta V_{IN\_MIN} \cdot N2}}$$

DC component of secondary current: $I_{2\_DC} = I_{out} \cdot D$.
RMS value of AC component of secondary current:

$$I_{2\_AC\_RMS} = \sqrt{I_{2\_RMS}^2 - I_{2\_DC}^2}$$

INDUCTOR L

Average current $I_L = I_{out}$;
Peak-to-peak inductor current ripple:

$$\Delta I_L = \frac{V_{out} \cdot (1 - \frac{V_{out} \cdot N1}{\eta \cdot V_{in} \cdot N2})}{L \cdot F}$$

where L – inductance in henry.
RMS value of inductor ripple current: $I_{L\_RMS} = \Delta I_L / \sqrt{12}$
The condition of CCM: $\Delta I_L / 2 < I_{out}$. If CCM is desired over entire input voltage range:

$$L_{MIN} > \frac{V_{out} \cdot (1 - V_{out} \cdot N1/(\eta V_{IN\_MAX} \cdot N2))}{2 \cdot F \cdot I_{out}}$$

In practice, L is selected several times larger than L$_{MIN}$, so that
ΔI$_L$≤(0.3...0.6)·Iout.

SWITCH Q
Peak current: I$_{Q\_PK}$=Iout·N2/N1+ΔI$_L$/2+Iμ
Center value of current pulse (midpoint of the ramp):
Icv=Iout·N2/N1+Iμ/2;
Peak voltage: V$_{Q\_PK}$=2·Vin (assuming N$_{DM}$=N1);
Conduction losses: P$_{Q\_COND}$≈(Iout·N2/N1)^2 · R$_{dson}$ ·D,
where R$_{dson}$ – drain to source resistance in on-state.

If PWM controller limits switch peak current to I$_{PK\_LIM}$, maximum achievable load current: I$_{OUT\_MAX}$=(I$_{Q\_PK}$-ΔI$_L$/2-Iμ)·N1/N2

OUTPUT RECTIFIERS D2, D3
Peak current: I$_{D2,3\_PK}$=Io+ΔI$_L$/2;
Average current: I$_{D2\_AV}$=Io·D, I$_{D3\_AV}$=Io·(1-D).
Peak plateau voltage: V$_{D2,3}$_pk=V$_{IN\_MAX}$·N2/N1.
Note. Instantaneous peak voltage may exceed the plateau by severalfold due to spikes caused by leakage inductance and reverse recovery current of the rectifiers. As a rule of thumb, we select the rectifiers with rated voltage 2-3 times greater than peak plateau voltage V$_{D2,3\_PK}$.

OUTPUT CAPACITOR
Peak current: Ic=±ΔI$_L$/2;
RMS current: Ic_rms=ΔI$_L$/(2·√3)
Net output capacitance is selected to satisfy ripple requirements (see Chapter 5.2) and transient response (see Chapter 5.3).

AUXILIARY CAPACITOR Caux
DC voltage: V$_{CAUX}$=Vin.

INPUT CURRENT
Average input current: I$_{IN\_AV}$=Vout·Iout/(η·Vin);
AC component of input current, which is current through input capacitor (not shown on the diagram): I$_{IN\_AC\_RMS}$ = I$_{1\_AC\_RMS}$

1.8 FORWARD CONVERTER VARIATIONS

TWO SWITCH FORWARD

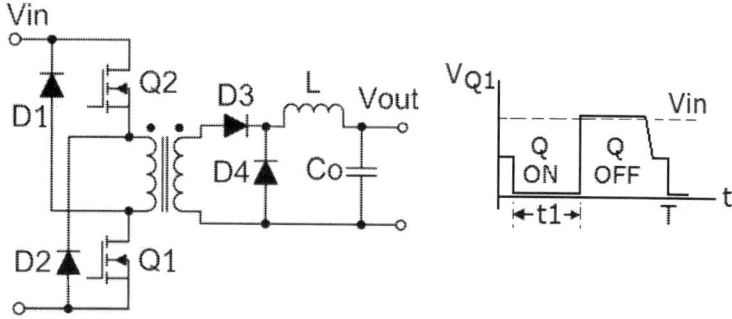

Q1 and Q2 turn ON and OFF at the same times. When both Q1 and Q2 are ON, rectifier D3 conducts and energy from input source is transferred to the output via inductor L. When Q1 and Q2 are OFF, inductor continues conducting via rectifier D4, while transformer core resets via D1 and D2. The equations are the same as for single-switch forward converter (see Chapter 1.7) except the peak MOSFETs voltage: $V_{Q\_PK}=V_{in}$ and duty cycle $D<0.5$.

ACTIVE CLAMP FORWARD

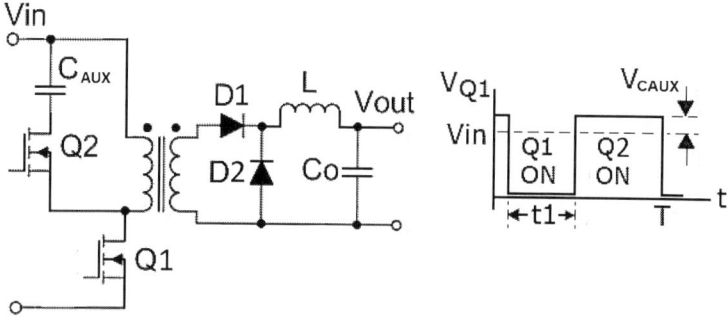

When Q1 is ON, Q2 is OFF, rectifier D1 conducts and energy from input source is transferred to the output via inductor L. When Q1 is OFF, Q2 is ON, inductor continues conducting via D2. During this time magnetizing current flows via Q2, which resets transformer core and charges auxiliary capacitor C_{AUX}. After the magnetizing current reaches zero it starts flowing in the opposite direction, sourced from the clamp capacitor Caux. When Q2 turns off, Q1 drain voltage (V_{Q1}) will fly down and Q1 may have zero voltage turn on.

Auxiliary capacitor voltage:
$$V_{CAUX} = \frac{Vin \cdot D}{1 - D}$$

The value of C_{AUX} is selected that resonant time with transformer magnetizing inductance is ten times greater than maximum off-time:

$$C_{AUX} = \frac{10 \cdot (1 - Dmin)^2}{L_\mu \cdot (2\pi \cdot F)^2}$$

where **Dmin** – minimum duty cycle at high line.

Peak MOSFETs voltage:
$$V_{Q\_PK} = Vin + V_{CAUX} = \frac{Vin}{1 - D} = \frac{\eta \cdot Vin^2}{\eta \cdot Vin - Vout \cdot N1/N2}$$

where D=t1/T - duty cycle, N1 and N2 – transformer's primary and secondary turns, η – converter's efficiency.

Rectifier D1 plateau voltage:
$$V_{D1\_PK} = \frac{Vin \cdot Vout}{\eta \cdot Vin - Vout \cdot N1/N2}$$

V_{D1_PK} reaches maximum at low line when V_{CAUX} is maximum. Other relationships are similar to that of single-switch forward converter (see Chapter 1.7). Duty cycle can be >0.5 (typically, designed to be around 0.7 at low line).

1.9 FLYBACK CONVERTER IN DCM

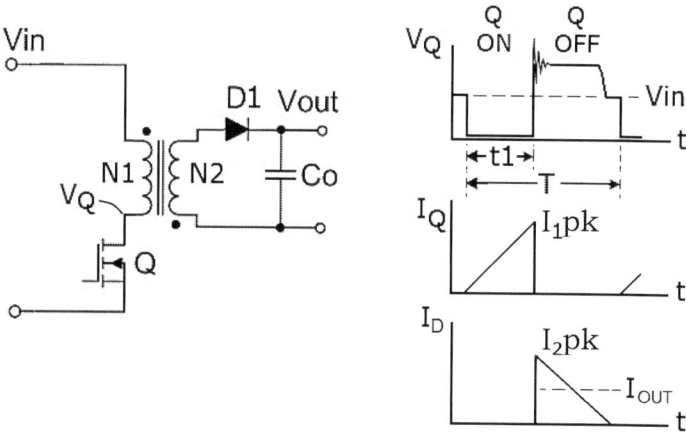

During ON state of the switch Q, energy is accumulated in the core of the transformer (which acts as an inductor), while output rectifier is reverse biased. During OFF state the transformer releases its entire stored energy. Most of the energy is released via the transformer's secondary into the load and output capacitor Co, the rest is dissipated in snubbers and primary clamp circuit (if any). Switch Q turns ON at zero current but its peak current I_{1PK} is higher than in other topologies.

OUTPUT POWER

In discontinuous conduction mode (DCM) from power balance:
$$Po = \eta \cdot L\mu \cdot I_1pk^2 \cdot F/2$$

where $Po = Iout \cdot Vout$ – output power; η - efficiency of the converter; $L\mu$ – transformer's magnetizing (primary) inductance in henry; F- frequency in hertz; I_{1PK}- peak primary current.
Peak primary current is controlled to regulate output voltage.

DUTY CYCLE
In steady-state fixed frequency operation:
$$D = \frac{\sqrt{2P_o \cdot L\mu \cdot F/\eta}}{Vin}$$

where D=t1/T (D<1).

OUTPUT VOLTAGE

In steady state: Vout=ηVin²·D²/(2Lμ·F·Iout).

TRANSFORMER

Peak primary current: $I_{1PK} = \sqrt{(2Po/\eta \cdot L\mu \cdot F)}$.
Peak secondary current: $I_{2PK} = I_{1pk} \cdot N1/N2$,
where N1/N2 - primary to secondary turns ratio.

If DCM is desired over entire input voltage range:

$$L\mu < \frac{\eta \cdot V_{IN\_MIN}^2 \cdot D_{LL}^2}{2Po \cdot F}$$

where D_{LL} – desired maximum duty cycle at low line (D_{LL} < Dmax, where Dmax – maximum duty cycle of PWM controller). In most controllers Dmax is 0.45 to 0.99. If Dmax is close to 1.0, it is common to set D_{LL} around 0.7 to limit maximum FET voltage. For selected Lμ, duty cycle at low line:

$$D_{LL} = \frac{\sqrt{2P_o \cdot L\mu \cdot F/\eta}}{V_{IN\_MIN}}$$

Turns ratio is usually selected such that at low line and full load the transformer will be near the boundary of DCM and CCM:

$$\frac{N1}{N2} \approx \frac{D_{LL} \cdot V_{IN\_MIN}}{Vout \cdot (1 - D_{LL})}$$

The core's reset time (the time it takes the transformer to completely demagnetize) is given by:

$$t_{RESET} = \frac{L\mu \cdot I_1 pk}{(N1/N2) \cdot Vout}$$

RMS value of primary current: $I_{1\_RMS} = I_{1PK} \cdot \sqrt{(D/3)}$.

It reaches maximum at low line:

$$I_{1\_RMS\_MAX} = \sqrt[4]{\frac{8 \cdot P_o^3}{9 \cdot L\mu \cdot F \cdot \eta^3 \cdot V_{IN\_MIN}^2}}$$

DC component of primary current: $I_{1\_DC} = V_{out} \cdot I_{out}/(\eta \cdot V_{IN})$

RMS value of AC component of primary current:

$$I_{1\_AC\_RMS} = \sqrt{I_{1\_RMS}^2 - I_{1\_DC}^2}$$

RMS value of secondary current:

$$I_{2\_RMS} = I_2pk \cdot \sqrt{\frac{t_{reset} \cdot F}{3}} = \sqrt[4]{\frac{8 \cdot P_o^3 \cdot T^2}{9 \cdot L\mu \cdot F \cdot \eta^3 \cdot V_{OUT}^2}}$$

DC component of secondary current: $I_{2\_DC} = I_{out}$.
RMS value of AC component of secondary current:

$$I_{2\_AC\_RMS} = \sqrt{I_{2\_RMS}^2 - I_{out}^2}$$

The required core and turns are determined by the **N1·Ac** product:

$$N1 \cdot Ac = \frac{L\mu \cdot I_1pk \cdot 10^8}{B_{PK}}$$

where **Ac** - core's equivalent cross-sectional area in sq.cm, **Bpk** – desired peak magnetic flux (Gauss).
See POWER INDUCTOR DESIGN (chapter 4.4) for selecting core size.
Since in flyback the magnetizing current is primary current, an air gap must be introduced in ferrite cores to prevent the magnetic material saturation:

$$lg = \frac{0.4 \cdot \pi \cdot N1 \cdot I_1pk}{B_{PK}} - \frac{lm}{\mu_r}$$

where **lg** – net length of air gap (cm), **lm** – effective magnetic core path length (cm), μ_r – relative permeability of ungapped core.

SWITCH Q
Peak current: $I_{Q\_PK} = I_{1PK}$;

Conduction losses: $P_{Q\_COND} = I_1rms^2 \cdot R_{dson}$, where R_{dson} - drain to source resistance of Q in on-state.
Peak plateau voltage: $V_{Q\_pk} = V_{IN\_MAX} + Vout \cdot N1/N2$.

Note. Instantaneous peak voltage can exceed the plateau by severalfold due to spikes caused by leakage inductance. Its actual value depends on the leakage inductance, primary snubber and clamp circuit (if any). As a rule of thumb, we select the MOSFET with rated voltage 2-3 greater than V_{Q_PK}. Otherwise we can use flyback either with two switches or with active clamp reset (see 1.11) that limits V_{Q_PK}.

If PWM controller limits switch peak current to I_{PK_LIM}, maximum achievable load current for given **Vout**:

$$I_{OUT\_MAX} = \frac{\eta \cdot L\mu \cdot F \cdot I_{PK\_LIM}^2}{2 \cdot Vout}$$

RECTIFIER D1

Peak current: $I_{D\_PK} = I_{2PK}$;
Average current: $I_D\_av = Iout$;
Reverse voltage plateau: $V_{D\_PK} = Vout + V_{IN\_MAX} \cdot N2/N1$.

Note. Instantaneous peak voltage can exceed the plateau by severalfold due to spikes caused by leakage inductance. As a rule of thumb, we select the diode with rated voltage 2-3 times greater than voltage plateau V_{D_PK}.

OUTPUT CAPACITOR Co

RMS current:

$$Ic\_rms = \sqrt{I_2rms^2 - Iout^2}$$

Net output capacitance is selected to satisfy ripple requirements (see Chapter 5.2) and transient response (see Chapter 5.3).

INPUT CURRENT

Average input current: $I_{IN\_AV} = Vout \cdot Iout/(\eta \cdot Vin)$;
AC component of input current, which is current through input capacitor (not shown on the diagram): $I_{IN\_AC\_RMS} = I_{1\_AC\_RMS}$.

1.10 FLYBACK CONVERTER IN CCM

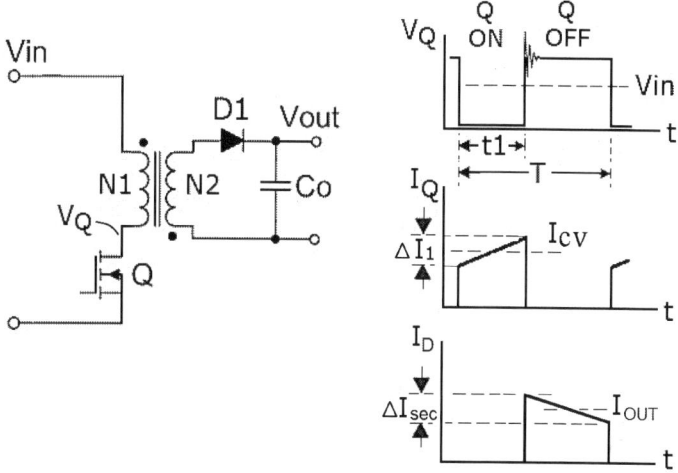

During ON state of the switch Q, energy is accumulated in the core of the transformer (which acts an inductor), while output rectifier is reverse biased. During OFF state a portion of stored energy is released via the transformer's secondary into the load and output capacitor Co. In continuous conduction mode (CCM) the magnetizing current does not reach zero.

OUTPUT VOLTAGE
In steady-state CCM:

$$Vout = \frac{\eta \cdot Vin \cdot N2}{N1} \cdot \frac{D}{1-D}$$

where η – converter efficiency; N1 and N2 - transformer's primary and secondary turns respectively; $D = t1/T$ - duty cycle ($D<1$).

DUTY CYCLE
In a steady state operation:

$$D = \frac{1}{1 + \frac{\eta V_{IN}}{Vout} \cdot \frac{N2}{N1}}$$

TRANSFORMER

Turns ratio must satisfy the inequation:

$$\frac{N2}{N1} \geq \frac{(V_{out} + V_{rect}) \cdot (1 - D_{MAX})}{\eta \cdot V_{IN\_MIN} \cdot D_{MAX}}$$

where D_{MAX} - maximum duty cycle of PWM (in most controllers $D_{MAX} = 0.45$ to 0.99); V_{rect} – rectifier voltage drop.

Note. If PWM controller has D_{max} close to 1.0, to lower peak voltage at the primary FET, it is common to select turns ratio such that duty cycle at low line is around 0.7-0.75:

$$\frac{N2}{N1} \approx \frac{0.4 \cdot (V_{out} + V_{rect})}{\eta V_{IN\_MIN}}$$

Peak to peak ramp of primary current: $\Delta I_1 = V_{in} \cdot D / (L\mu \cdot F)$, where $L\mu$ – magnetizing (primary) inductance,

Center value of primary current pulse (midpoint of the ramp): $I_{cv} = P_o / (\eta V_{in} \cdot D)$, where $P_o = V_{out} \cdot I_{out}$ – output power.
Peak primary current: $I_{1PK} = I_{cv} + \Delta I_1 / 2$

The condition of CCM: $\Delta I_1 / 2 < I_{cv}$, which requires:

$$L\mu > \frac{\eta \cdot V_{IN}^2 \cdot D^2}{2 P_o \cdot F}$$

where F- frequency is hertz.
If CCM is desired over entire input voltage range:

$$L\mu \geq \frac{\eta \cdot V_{IN\_MAX}^2}{2 P_o \cdot F \cdot (1 + \frac{\eta \cdot V_{IN\_MAX}}{V_{out}} \cdot \frac{N2}{N1})^2}$$

where V_{IN_MAX}- maximum input voltage.
In practice, $L\mu$ is selected $1.4\ldots 2$ times larger in order to keep CCM at $50\ldots 70\%$ of rated power P_o.

RMS value of primary current:

$$I_1\text{rms} = \sqrt{D \cdot (Icv^2 + \Delta I_1^2/12)}$$

Maximum RMS of primary current appears at low line.
If $\Delta I_1/3 \ll Icv$:

$$I_{1\_RMS\_MAX} \approx Icv \cdot \sqrt{D} = \frac{Po}{\eta V_{IN\_MIN} \cdot \sqrt{1 + \frac{\eta V_{IN\_MIN}}{Vout} \cdot \frac{N2}{N1}}}$$

DC component of primary coil current: $I_{1\_DC} = Po/(\eta Vin)$.

RMS value of AC component the current in primary coil, which determines AC copper losses:

$$I_{1\_AC\_RMS} = \sqrt{I_{1\_RMS}^2 - I_{1\_DC}^2} \approx Iout \cdot \sqrt{\frac{N2 \cdot Vout}{N1 \cdot \eta \cdot V_{IN\_MIN}}}$$

Center value of secondary current pulse (midpoint of the ramp): $Isec\_cv = Iout/(1-D)$.

Peak to peak secondary current ramp:

$$\Delta Isec = \frac{Vout}{L} \cdot \frac{N1^2}{N2^2} \cdot \frac{1-D}{F}$$

RMS value of secondary current:

$$I_2\text{rms} = \sqrt{\frac{Iout^2}{1-D} + \frac{\Delta Isec^2}{12} \cdot (1-D)}$$

If $\Delta Isec \ll Iout$:

$$I_2\text{rms} \approx Iout \cdot \sqrt{1 + \frac{N1 \cdot Vout}{N2 \cdot \eta V_{IN\_MIN}}}$$

DC component of current in secondary coil: $I_{2\_DC} = Iout$.
RMS value of AC component of current in secondary coil, which determines AC copper losses:

$$I_{2\_AC\_RMS} \approx Iout \cdot \sqrt{\frac{N1 \cdot Vout}{N2 \cdot \eta V_{IN\_MIN}}}$$

The required core and turns are determined by the **N1·Ac** product:

$$N1 \cdot Ac = \frac{L\mu \cdot I_1 pk \cdot 10^8}{B_{PK}}$$

where **Ac** - core's equivalent cross-sectional area in sq.cm, **Bpk** – desired peak magnetic flux (Gauss).
See POWER INDUCTOR DESIGN (chapter 4.4) for selecting core size.
Since in flyback the magnetizing current is primary current, in ferrite cores an air gap must be introduced to prevent the magnetic material saturation:

$$lg = \frac{0.4 \cdot \pi \cdot N1 \cdot I_1 pk}{B_{PK}} - \frac{lm}{\mu_r}$$

where **lg** – net length of air gap (cm), **lm** – effective magnetic core path length (cm), μ_r – relative permeability of ungapped core.

SWITCH Q

Peak current: $I_{Q\_PK} = Icv + \Delta I_1/2$.
Maximum plateau voltage: $V_{Q\_PK} = V_{IN\_MAX} + N1 \cdot Vout/N2$.

Note. Instantaneous peak voltage can be severalfold higher than the plateau due to spikes caused by leakage inductance. Its actual value depends on the leakage inductance, primary snubber and clamp circuits (if any). As a rule of thumb, we select the MOSFET with rated voltage 2-3 times greater than Peak plateau voltage V_{Q_PK}, or use flyback either with two switches or with active clamp reset (see 1.11).

Conduction losses: $P_{Q\_COND} = I_1 rms^2 \cdot R_{dson}$,

where R_{dson} - drain to source resistance of Q in on-state.

If PWM controller limits switch peak current to I_{PK_LIM}, maximum achievable load current for given **Vout**:

$$I_{OUT\_MAX} = \frac{I_{PK\_LIM} - \frac{0.5 \cdot V_{IN\_MIN} \cdot D_{LL}}{L \cdot F}}{Vout} \cdot (\eta \cdot V_{IN\_MIN} \cdot D_{LL})$$

where D_{LL} – duty cycle at low line:

$$D_{LL} = \frac{1}{1 + \frac{\eta V_{IN\_MIN}}{Vout} \cdot \frac{N2}{N1}}$$

RECTIFIER D1

Average current: I_{D_AV} = Iout.
Peak current: $I_{D\_PK} = I_1 pk \cdot N1/N2$.
Peak plateau of the voltage pulse: $V_{D\_PK} = Vout + V_{IN\_MAX} \cdot N2/N1$.

Note. Instantaneous peak voltage can exceed the plateau by severalfold due to spikes caused by leakage inductance and reverse recovery current of the diode. As a rule of thumb, select the diodes with rated voltage 2-3 times greater than plateau voltage V_{D_PK}.

OUTPUT CAPACITOR Co

RMS current:

$$Ic_{rms} = \sqrt{I_2 rms^2 - Iout^2} = \sqrt{\frac{Iout^2 \cdot D}{1-D} + \Delta Isec^2 \cdot \frac{1-D}{12} - Iout^2}$$

If $\Delta Isec \ll Iout$, maximum rms current which appears at low line:

$$Ic_{rms} \approx Iout \cdot \sqrt{\frac{D}{1-D}} = Iout \cdot \sqrt{\frac{Vo \cdot N1}{\eta V_{IN\_MIN} \cdot N2}}$$

Net output capacitance is selected to satisfy ripple requirements (see Chapter 5.2) and transient response (see Chapter 5.3).

INPUT CURRENT

Average input current: $I_{IN\_AV} = Vout \cdot Iout/(\eta \cdot Vin)$;
AC component of input current, which is current through input capacitor (not shown on the diagram): $I_{IN\_AC\_RMS} = I_{1\_AC\_RMS}$.

1.11 FLYBACK CONVERTER VARIATIONS

TWO SWITCH FLYBACK

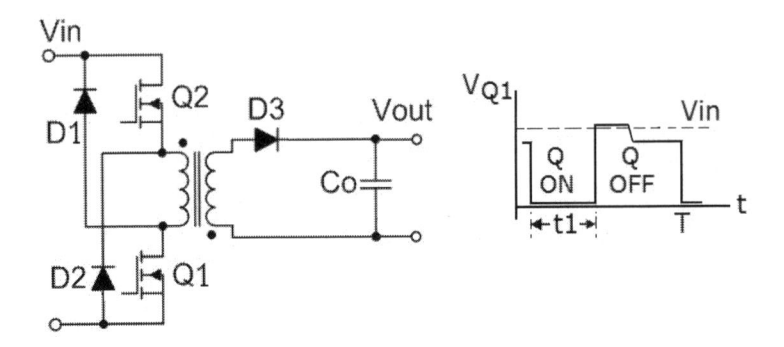

Q1 and Q2 are driven synchronously. When they are ON, energy is accumulated in the transformer's core as in a classical flyback. During OFF state of Q1, Q2 the stored energy is released via the transformer's secondary into the load and output capacitor Co. The energy associated with leakage inductance is returned to the input source via D1, D2. Duty cycle must be $D<0.5$. Turns ratio:

$$\frac{N2}{N1} \geq \frac{(V_{out} + V_{rect})}{\eta \cdot V_{IN\_MIN}}$$

Other equations are similar to that of single-switch flyback converters (see Chapter 1.9, 1.10) except the peak MOSFETs voltage: $V_{Q\_PK} = V_{in}$.

ACTIVE CLAMP FLYBACK

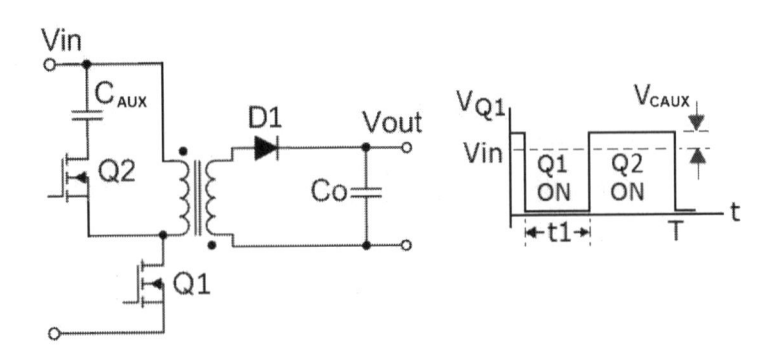

When Q1 is ON, Q2 and D1 are OFF, and energy is accumulated in the transformer's core like in a classical flyback. When Q1 is OFF, Q2 is ON, most of the stored energy is released via the transformer's secondary into the load and output capacitor Co. A portion of magnetizing current flows through Q2 and auxiliary capacitor C_{AUX}. After the magnetizing current reaches zero, it starts flowing in the opposite direction via Q2, sourced from C_{AUX}.

Auxiliary capacitor voltage in CCM:

$$V_{C\_AUX} = \frac{Vin \cdot D}{1 - D} = \frac{Vout \cdot N1}{\eta \cdot N2}$$

where $D = t1/T$ - duty cycle; N1 and N2 – transformer's primary and secondary turns, η – converter's efficiency.

The value of C_{AUX} is selected so that resonant time with transformer magnetizing inductance is ten times greater than maximum off-time:

$$C_{AUX} = \frac{10 \cdot (1 - Dmin)^2}{L_\mu \cdot (2\pi \cdot F)^2}$$

where Dmin – minimum duty cycle at high line:

$$Dmin = \frac{1}{\frac{\eta V_{IN\_MAX} \cdot N2}{Vout \cdot N1} + 1}$$

Peak MOSFET's voltage:

$$V_{Q\_PK} = Vin + V_{C\_AUX} = Vin + \frac{Vout \cdot N1}{\eta \cdot N2}$$

Other relationships are similar to that of classical single-switch flyback converter (see Chapter 1.9).

1.12 HALF BRIDGE

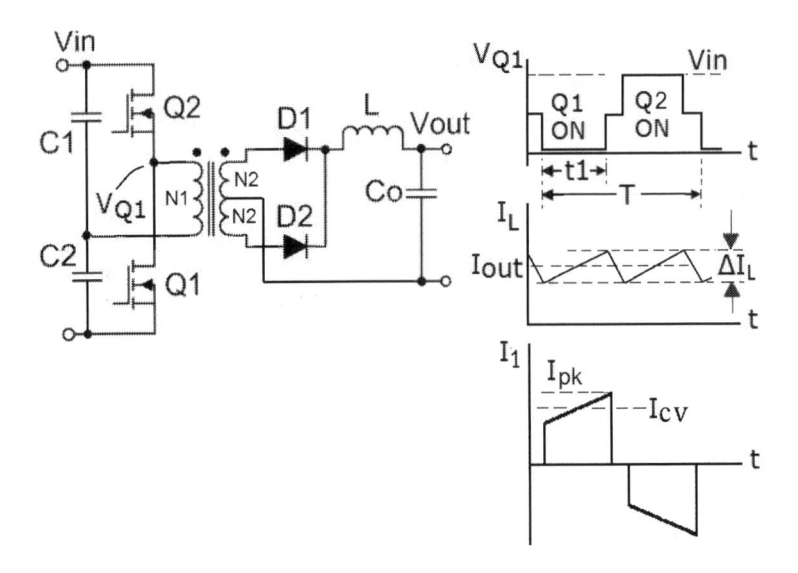

Capacitors C1 and C2 create voltage at their center point equal to half the input voltage **Vin**. When Q1 is ON (time interval **t1**), Vin/2 is applied across the primary winding and diode D2 conducts. During this time energy from input source is transferred to the output via inductor L and in the process, some energy is also accumulated in the inductor and in transformer's core. During "dead time" inductor continue conducting load current via both diodes D1 and D2. When Q2 is on, the same process repeats with diode D1.

OUTPUT VOLTAGE
When inductor L operates in continuous conduction mode:

$$Vout = \eta Vin \cdot D \cdot N2/N1$$

where η - efficiency of the converter; D - duty cycle of the switches Q1, Q2 (D<0.5); N1- primary turns of the transformer, N2- turns of each half-secondary.

DUTY CYCLE
In a steady state operation:

$$D \approx \frac{V_{out} \cdot N1}{\eta V_{in} \cdot N2}$$

TRANSFORMER
Turns ratio must satisfy the inequation:

$$N2/N1 \geq (V_{out}+V_{rect})/(\eta V_{IN\_MIN} \cdot D_{MAX}),$$

where D_{MAX}- maximum duty cycle of PWM, $D_{MAX} < 0.5$;
V_{IN_MIN} – minimum input voltage, **Vrect** – rectifier voltage drop.

To achieve **Dmax** at V_{IN_MIN}, minimum required **N1·Ac** product:

$$N1 \cdot Ac \geq \frac{V_{IN\_MIN} \cdot D_{max}}{4 \cdot F \cdot B_{pk} \cdot 10^{-8}}$$

where **Ac**- core cross-sectional area (in sq.cm); **Bpk** - peak working magnetic flux density in gauss (usually selected at 2000-3000 G to be below 0.7 of saturation flux); F - frequency is hertz.

Note. With such **N1·Ac** selection, magnetic flux can exceed **Bpk** during transients. The circuit should employ pulse-by-pulse current limit. For a conservative design, in the numerator of above equation for **N1·Ac** put V_{IN_MAX} instead of V_{IN_MIN}.

Magnetizing current: $I\mu = V_{in} \cdot D/2F \cdot L\mu$, where $L\mu$- transformer magnetizing inductance, D- duty cycle.
Maximum magnetizing current in steady state operation:
$$I\mu_{\_PK} = V_{IN\_MIN}/4F \cdot L\mu$$

Maximum RMS value of primary current (neglecting $I\mu$ and inductor ripple current ΔI_L):

$$I_1 rms = I_{out} \cdot \sqrt{\frac{2 \cdot V_{out} \cdot N2}{\eta V_{IN\_MIN} \cdot N1}}$$

Maximum RMS value of the current in each secondary:

$$I_2\text{rms} = \text{Iout} \cdot \sqrt{\frac{V_{out} \cdot N1}{\eta V_{IN\_MIN} \cdot N2}}$$

INDUCTOR L
Average current $I_L = \text{Iout}$;
Peak-to-peak inductor current ripple:

$$\Delta I_L = V_{out} \cdot (1 - D)/(L \cdot 2F)$$

RMS value of AC component of inductor current: $I_L\_\text{rms} = \Delta I_L/2\sqrt{3}$. The condition of continuous conduction mode: $\Delta I_L/2 < \text{Iout}$. This requires:

$$L_{MIN} > \frac{V_{out} \cdot \left(1 - \frac{V_{out} \cdot N1}{\eta V_{IN\_MAX} \cdot N2}\right)}{4 \cdot L \cdot F \cdot \text{Iout}}$$

In practice, L is selected several times larger than L_{MIN}, so that $\Delta I_L < (0.3...0.6) \cdot \text{Iout}$.

SWITCHES Q1, Q2
Peak voltage: $V_{Q\_PK} = V_{in}$;
Peak current: $I_{Q\_PK} = \text{Iout} \cdot N2/N1 + \Delta I_L/2 + I\mu$;
Center value of current pulse (midpoint of the ramp):
$I_{cv} = \text{Iout} \cdot N2/N1 + I\mu/2$;
RMS current in each FET:

$$I_Q\text{rms} = \frac{I_1\text{rms}}{\sqrt{2}} = \text{Iout} \cdot \sqrt{\frac{V_{out} \cdot N2}{\eta V_{IN\_MIN} \cdot N1}}$$

Conduction losses in each FET: $P_{cond} = I_Q\text{rms}^2 \cdot R_{dson}$, where R_{dson} - drain to source resistance of Q in on-state.

If PWM controller limits peak current of the switch to I_{PK_LIM}, maximum achievable load current: $\text{Iout}\_\text{MAX} = (I_{PK\_LIM} - \Delta I_L/2 - I\mu) \cdot N1/N2$

OUTPUT RECTIFIERS D1, D2
Peak current: $I_{D\_PK}=Iout+\Delta I_L/2$;
Average current: $I_{D\_AV}=Iout/2$.
Peak plateau voltage: $V_{D\_PK}=V_{IN\_MAX}\cdot N2/N1$.

Note. Instantaneous peak voltage can exceed the plateau by severalfold due to spikes caused by leakage inductance and reverse recovery current of the rectifiers. As a rule of thumb, we select the diodes with rated voltage 2-3 times greater than plateau V_{D_PK}.

INPUT CAPACITORS C1, C2
RMS current in each capacitor: $I_{CIN\_RMS}=I_{Q\_RMS}$
Maximum voltage: $V_{CIN\_MAX}=V_{IN\_MAX}/2$.

OUTPUT CAPACITOR
Peak current: $I_{C\_PK}=\Delta I_L/2$;
RMS current: $I_{C\_rms}=\Delta I_L/(2\cdot\sqrt{3})$.
Net output capacitance is selected to satisfy ripple requirements (see Chapter 5.2) and transient response (see Chapter 5.3).

INPUT CURRENT
Average input current: $I_{IN\_AV}=V_{out}\cdot I_{out}/(\eta V_{IN\_MIN})$.
AC component of input current is mostly absorbed by input capacitors C1, C2.

1.13 LLC HALF BRIDGE

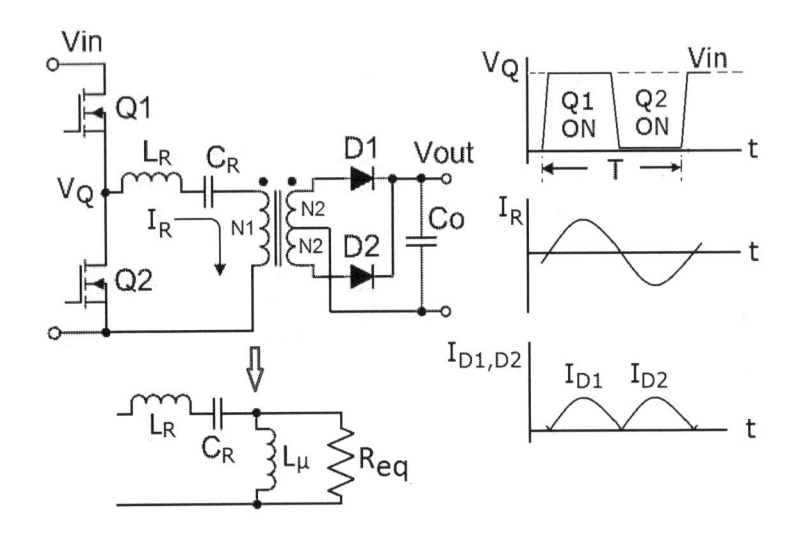

Two power switches Q1 and Q2 are driven complimentary with nearly 50% duty cycle each (less some small dead time) at variable frequency. The resonant tank L_R, C_R and acts as a voltage divider with the load and transformer's magnetizing inductance $L\mu$. The amount of energy delivered to the load dependents on the $L_R C_R$ impedance at a given frequency. The converter is designed to operate near resonance frequency at nominal input and rated load. It will then operate above resonance at high line and decreased load, and below resonance at low line. The component stresses below are given for operation at resonance.

OUTPUT VOLTAGE

In general:

$$V_{out}(\omega) = \frac{\eta V_{in} \cdot N2}{2 \cdot N1} \left| \frac{(j\omega L\mu) || R_{eq}}{(j\omega L\mu) || R_{eq} + j\omega L_R + 1/(j\omega C_R)} \right|$$

where η - efficiency of the converter; $L\mu$ - magnetizing inductance (henry); $\omega = 2\pi F$ - angular frequency (radians per second) corresponding to switching frequency $F = 1/T$; R_{eq} - equivalent load resistance reflected to the primary for fundamental harmonic of input pulses. In steady-state continuous conduction mode of L_R:

$$Req = \frac{8 \cdot (N1/N2)^2}{\pi^2} \cdot \frac{Vout}{Iout}$$

Resonant frequency in hertz:

$$Fr = \frac{1}{2\pi \cdot \sqrt{L_R \cdot C_R}}$$

At resonance frequency (F=Fr) $L_R C_R$ impedance is zero:

$$Vout \approx \frac{\eta \cdot Vin \cdot N2}{2 \cdot N1}$$

TRANSFORMER

Required turns ratio for operation at resonance at nominal input Vin:

$$N1/N2 = \eta \cdot Vin / (2 \cdot Vout + Vrect)$$

where **Vrect** – rectifier voltage drop.
Minimum required **N1·Ac** product:

$$N1 \cdot Ac \geq \frac{0.5 \cdot V_{IN}}{4 \cdot Fr \cdot Bpk \cdot 10^{-8}}$$

where **Ac** - core cross-sectional area (in sq.cm); **Bpk** - peak magnetic flux density in gauss (usually selected around 2000-3000 G).
RMS value of primary current (neglecting magnetizing current):

$$I_1 rms \approx \frac{\pi \cdot Iout \cdot N2}{2\sqrt{2} \cdot \eta \cdot N1}$$

RMS value of current in each half secondary:

$$I_2 rms \approx \frac{\pi \cdot Iout}{4}$$

DC component of current in each half secondary:

$$I_{2\_DC} \approx \frac{Iout}{2}$$

RESONANT INDUCTOR L$_R$

At high line and no-load F>>Fr. Then, neglecting effects of C$_R$:

$$Vout \approx \frac{L\mu}{L\mu + L_R} \cdot \frac{\eta \cdot V_{IN} \cdot N2}{2 \cdot N1}$$

The L$_R$ is selected to regulate Vout at high line and no load:

$$L_R > L\mu \cdot \left(\frac{\eta \cdot V_{IN\_MAX} \cdot N2}{2 \cdot Vout \cdot N1} - 1\right)$$

For recommended turns ratio N1/N2= $\eta \cdot$V$_{IN}$ /(2·Vout) :

$$L_R > L\mu \cdot \left(\frac{V_{IN\_MAX}}{V_{IN}} - 1\right)$$

Peak L$_R$, C$_R$ current:

$$I_{LC\_PK} \approx \frac{\pi \cdot Iout \cdot N2}{2 \cdot \eta \cdot N1}$$

RMS current through L$_R$, C$_R$: I$_{LC\_RMS}$=I$_1$rms .
If PWM controller limits peak LC current to I$_{PK\_LIM}$, maximum load current: I$_{OUT\_MAX}$=2$\eta \cdot$N1· I$_{PK\_LIM}$/($\pi \cdot$N2).

RESONANT CAPACITOR C$_R$

For desired resonant frequency Fr:

$$C_R = \frac{1}{(2\pi \cdot Fr)^2 \cdot L_R}$$

Peak voltage across C$_R$ at nominal input:

$$V_{C\_PK} = \frac{V_{IN}}{2} + \frac{Iout \cdot N2}{4 \cdot F \cdot N1}$$

QUALITY FACTOR

At low input the resonant tank has to provide maximum gain M_{MAX} to support **Vout**:

$$M_{MAX} = \frac{2 \cdot Vout \cdot N1}{\eta \cdot V_{IN\_MIN} \cdot N2}$$

For recommended turns ratio $N1/N2 = \eta \cdot V_{IN}/(2 \cdot Vout)$:

$$M_{MAX} = V_{IN}/V_{IN\_MIN}$$

The gain of resonant circuit:

$$M_G(\omega) = \left| \frac{(j\omega L\mu) || Req}{(j\omega L\mu) || Req + j\omega L_R + 1/(j\omega C_R)} \right|$$

where $\omega = 2\pi \cdot F$ - angular frequency.
Rearranging the above equation gives:

$$M_G(F_n) = \frac{1}{\sqrt{(1 - \frac{\lambda}{F_n^2} + \lambda)^2 + Q^2 \cdot (\frac{1}{F_n} + F_n)^2}}$$

where: $Q = \frac{\sqrt{L_R/C_R}}{Req}$ - quality factor; $F_n = F/F_R$ - normalized frequency; $\lambda = L_R/L\mu$ - inductance ratio.

Achieving regulation at low line at certain selected minimum frequency F_{MIN} requires:

$$M_G(F_{MIN}) \geq M_{MAX}$$

If calculated L_R and C_R don't provide M_{MAX} at $F \geq F_{MIN}$ we need to lower quality factor Q of the resonant tank to boost its gain below resonant frequency. You may choose Q from a family of plots of the gain versus normalized frequency ([1]). A typical range is $Q = 0.2...0.5$

at rated load. A few iterations are usually needed to optimize the design.

For a typical application when input voltage varies +/-10% from nominal, we need $L_R > 0.1 \cdot L_\mu$ to provide $M_{MAX} = 1.1$ at V_{IN_MIN}. For such application we can choose for example $L_R = 0.12 \cdot L_\mu$. Below are the plots of resonant circuit gain M_G versus normalized frequency F_n for $\lambda = 0.12$. For such inductance ratio we get sufficient gain at $Q \leq 0.4$ and frequencies $0.3...0.5 \cdot F_R$.

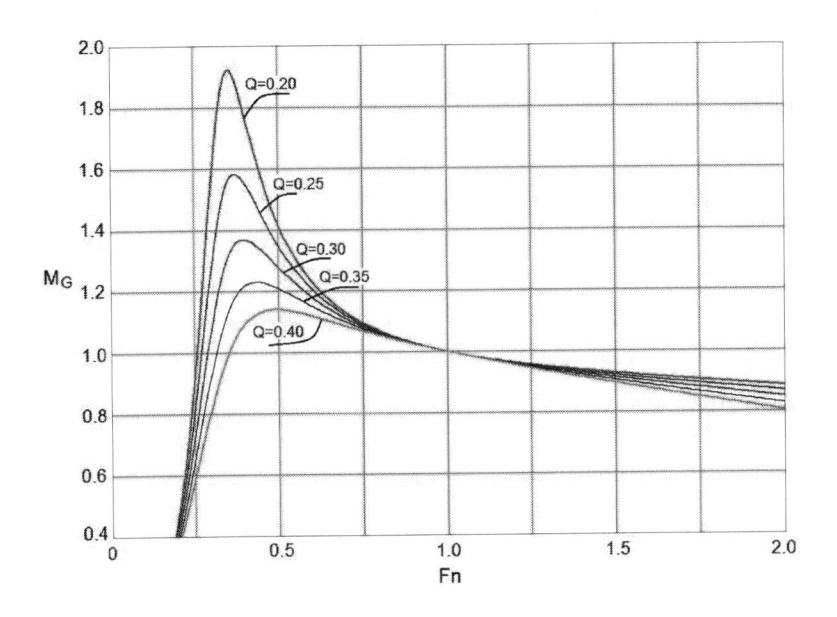

SWITCHES Q1, Q2

Peak voltage: $V_{Q\_PK} = V_{in}$;

Peak current: $I_{Q\_PK} = I_{LC\_PK}$;

RMS current in each switch: $I_{Q\_RMS} = I_{1rms}/\sqrt{2}$.

Conduction losses in each MOSFET: $P_{cond} = I_{Qrms}^2 \cdot R_{dson}$, where R_{dson} - drain to source resistance in on-state.

Condition of zero-voltage switching:
$$T_{DEAD} = 16 \cdot (2 \cdot C_{DS} + C_{STRAY}) \cdot F \cdot L_\mu,$$
where C_{DS} - drain to source capacitance of each switch, C_{STRAY} - total stray capacitance at switches junction.

OUTPUT RECTIFIERS D1, D2
Peak current: $I_{D\_PK}=0.5\pi \cdot Iout$;
Average current: $I_{D\_AV}=Iout/2$.
Peak voltage: $V_{D\_PK}=V_{IN} \cdot N2/N1$.

OUTPUT CAPACITOR
Peak current: $I_{C\_PK}= I_{D\_PK}$;
RMS current:
$$I_{C\_RMS} = Iout \cdot \sqrt{0.125 \cdot \pi^2 - 1} \approx 0.48 \cdot Iout$$
Net output capacitance is selected to satisfy ripple requirements (see Chapter 5.2) and transient response (see Chapter 5.3).

INPUT CURRENT
Average input current: $I_{IN\_AV}=Vout \cdot Iout/(\eta V_{IN\_MIN})$.
RMS value of AC component of input current, which is current through input capacitor (not shown on diagram):

$$I_{IN\_AC\_RMS} = \sqrt{I_{Q\_RMS}^2 - I_{IN\_AV}^2} \approx \frac{Iout \cdot Vout}{\eta \cdot V_{IN}} \cdot \sqrt{\frac{\pi^2}{4\eta^2} - 1}$$

1.14 PHASE SHIFTED FULL BRIDGE WITH CURRENT DOUBLER

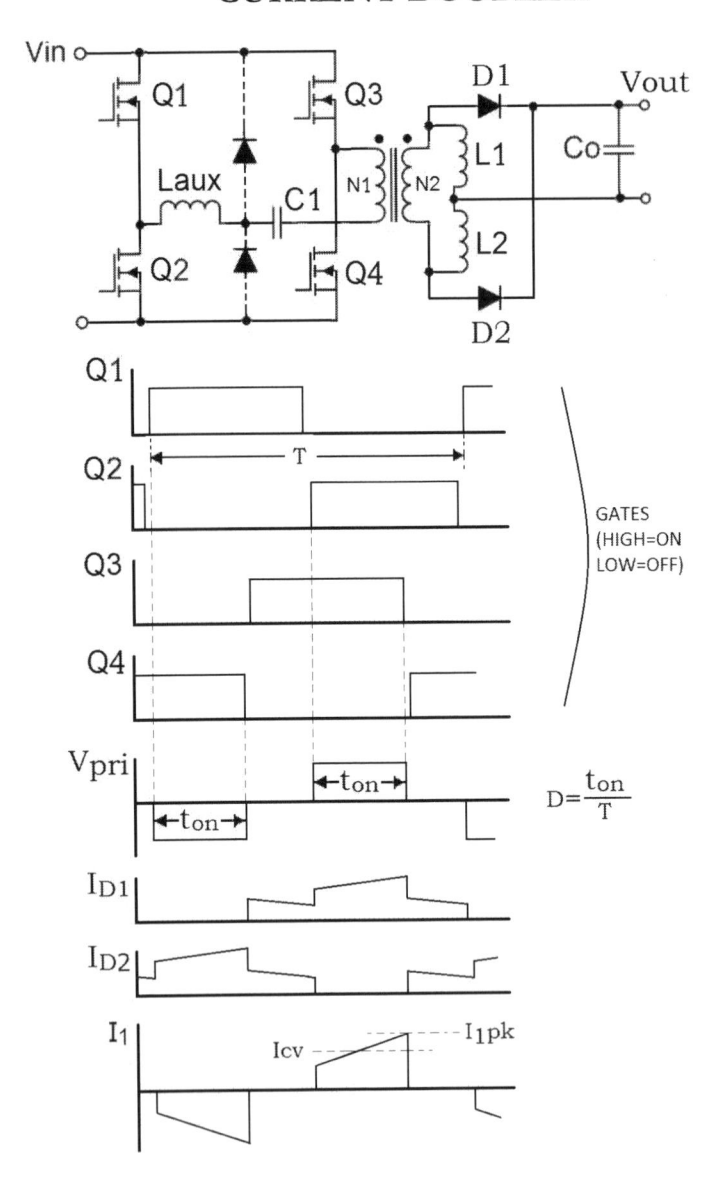

$$D = \frac{t_{on}}{T}$$

The converter has four power switches Q1-Q4. Switches in each vertical pair are driven complimentary with almost 50% duty cycle (less some small dead time). Power is transferred when both diagonal switches are ON (t_ON time interval). The effective transformer's duty cycle D_EFF is determined by the phase shift between these diagonal switches ON times. Such converter is usually implemented with current doubler rectifier, which has two inductors L1 and L2, each carrying half the total load current. Current doubler eliminates the need for center tapping in transformer's secondary and reduces ripple current through the output capacitors. An optional auxiliary inductor Laux facilitates zero voltage switching. An auxiliary capacitor C1 removes DC offset of magnetizing current.

OUTPUT VOLTAGE

When output inductors operate in continuous conduction mode:

$$V_{out} = \frac{\eta V_{in} \cdot D_{EFF} \cdot N2}{N1} \cdot \frac{L\mu}{L\mu + L_{aux}}$$

where η - efficiency of the converter; D_{EFF} – effective transformer's duty cycle ($D_{EFF} < 0.5$); N1 and N2 - primary and secondary turns, $L\mu$ - magnetizing (primary) inductance; L_{aux} – auxiliary inductance.

EFFECTIVE DUTY CYCLE

In a steady state operation effective duty cycle of transformer:

$$D_{EFF} = \frac{V_{out} \cdot N1}{\eta \cdot V_{in} \cdot N2} \cdot \frac{L\mu + L_{aux}}{L\mu}$$

INDUCTORS L1, L2

Average current $I_{L\_AV} = I_{out}/2$;
Peak-to-peak inductor current ripple: $\Delta I_L = V_{out} \cdot (1 - D_{EFF})/(L \cdot F)$.

The condition of continuous conduction mode: $\Delta I_L < 2 \cdot I_{L\_AV}$. This requires:

$$L_{MIN} > \frac{V_{out} \cdot (1 - \frac{V_{out} \cdot N1}{\eta V_{INMAX} \cdot N2})}{L \cdot F \cdot I_{out}}$$

In practice, L is selected several times greater than L_{MIN}, such that $\Delta I_L \leq (0.3-0.6) \cdot I_{out}$.

TRANSFORMER

Turns ratio must satisfy the inequation:

$$\frac{N2}{N1} \geq \frac{Vout + Vrect}{\eta V_{IN\_MIN} \cdot Dmax} \cdot \frac{L\mu + Laux}{L\mu} \approx \frac{2Vout}{\eta V_{IN\_MIN}}$$

where **Dmax** - maximum effective duty cycle of PWM (Dmax<0.5); V_{IN_MIN} – minimum input voltage, **Vrect** – rectifier voltage drop.

Minimum required **N1·Ac** product, where Ac- core cross-sectional area (in sq.cm):

$$N1 \cdot Ac \geq \frac{V_{IN\_MIN} \cdot Dmax}{2 \cdot F \cdot Bpk \cdot 10^{-8}} \approx \frac{V_{IN\_MIN}}{4 \cdot F \cdot Bpk \cdot 10^{-8}}$$

where **Bpk** - peak working magnetic flux density in gauss (usually selected at 2000-3000 G); F- frequency is hertz.

Note. With such **N1·Ac** selection, magnetic flux can exceed **Bpk** during transients. The circuit should employ pulse-by-pulse current limit. For a conservative design, in the numerator of above equation for **N1·Ac** put V_{IN_MAX} instead of V_{IN_MIN}.

Peak magnetizing current: $I\mu = Vin \cdot D_{EFF}/F \cdot L\mu$.
Center value of each primary current pulse (midpoint of the ramp): $Icv = 0.5 \cdot Iout \cdot N2/N1$.
RMS value of primary current (neglecting ramp ΔI):

$$I_1 rms \approx Icv \cdot \sqrt{2D_{EFF}} = Iout \cdot \sqrt{\frac{Vout \cdot N2}{2 \cdot \eta \cdot Vin \cdot N1}}$$

RMS vale of secondary current (neglecting inductor's current ripple):

$$I_2 rms \approx 0.5 Iout \cdot \sqrt{2D_{EFF}} = Iout \cdot \sqrt{\frac{Vout \cdot N1}{2 \cdot \eta \cdot Vin \cdot N2}}$$

SWITCHES Q1-Q4

Peak to peak value of the ramp of current pulse:

$$\Delta I_Q = \frac{Vin \cdot D_{EFF}}{F \cdot L\mu} + \frac{\Delta I_L \cdot N2}{N1} + I\mu$$

Peak current: $I_{Q\_PK} = Icv + 0.5 \cdot \Delta I_Q$.
RMS current in each switch (neglecting ΔI_Q):

$$I_Q rms \approx Icv \cdot \sqrt{D_{EFF}} = \frac{Iout}{2} \cdot \sqrt{\frac{Vout \cdot N2}{\eta \cdot Vin \cdot N1}}$$

Peak voltage: $V_{Q\_PK} = Vin$.
Conduction losses in each FET: $Pcond = I_{Q\_RMS}^2 \cdot R_{dson}$,
where R_{dson} - drain to source resistance of Q in on-state.

OUTPUT RECTIFIERS D1, D2

Peak plateau of the voltage pulse: $V_{D\_PK} = V_{IN\_MAX} \cdot N2/N1$.
Peak current: $I_{D\_PK} = Iout + \Delta I_L/2$;
Average current $I_{D\_AV} = Iout/2$;
RMS current in each rectifier (neglecting ΔI_L):

$$I_{D\_RMS} = Iout \cdot \sqrt{0.5 \cdot \frac{Vout \cdot N1}{\eta \cdot Vin \cdot N2} + 0.25}$$

OUTPUT CAPACITOR Co

Peak to peak current ripple:

$$\Delta I_{C\_PKPK} = \frac{2 \cdot Vout}{L \cdot F} \cdot (0.5 - \frac{Vout \cdot N1}{\eta \cdot Vin \cdot N2})$$

RMS current: $I_{C\_RMS} = 0.5 \cdot \Delta I_{C\_PKPK}/\sqrt{3}$.
Net output capacitance is selected to satisfy ripple requirements (see Chapter 5.2) and transient response (see Chapter 5.3).

INPUT CURRENT

Average input current: $I_{IN\_AV} = Vout \cdot Iout/(\eta V_{IN})$.
RMS value of AC component of input current, which is current through input capacitor (not shown on diagram): $I_{IN\_AC\_RMS} = I_1 rms$.

1.15 TOPOLOGY SELECTION

Typical relative rankings of isolating DC-DC converter topologies:

	SWITCHING LOSSES	OVERALL EFFICIENCY	TRANSFORMER SIZE	PARTS COUNT	SWITCH UTILIZATION
Flyback DCM	Low	Low	Largest	Lowest	Lowest
Flyback CCM	Average	Low	Large	Lowest	Low
Single switch forward	Average	Average	Average	Low	Average
Active clamp forward	Low	High	Average	Average	High
Half-bridge	Average	Average	Smallest	Average	High
ZVT full bridge	Lowest	Highest	Small	Highest	Average
LLC	Lowest	Highest	Small	Average	Average

Suitable topologies for off-line low-voltage isolated DC power supply (3.3V<Vout<24V) with typical output voltage and current requirements.

	0-50W	50-100W	100-400W	400-1000W	1000-3000W
Flyback DCM	✓				
Flyback CCM	✓	✓			
Two switch flyback	✓	✓			
One switch forward	✓	✓			
Two switch forward		✓	✓		
Active clamp forward		✓	✓		
Half bridge			✓	✓	
LLC			✓	✓	
Phase shifted ZVT bridge				✓	✓

Above 3000W, I would use multiple interleaved phase shifted ZVT bridge converters to spread the losses. For low-voltage outputs (<3.3V) a two-stage conversion may be used: a pre-regulated bus (5...12V) followed by a buck converter.

2 POWER FACTOR CORRECTION

2.1 POWER FACTOR BASICS

Power factor is defined as ratio between real power P (watts) and apparent power S (a.k.a total power or volt-amps):

$$PF \stackrel{def}{=} P/S = P/(V_{RMS} \cdot I_{RMS})$$

where V_{RMS} and I_{RMS} – rms values of voltage and current.

Power factor is a property of the load on the electric source, rather than a property of the source. If a load draws current $i(t)$ from a source with periodic voltage $v(t)$, power factor in general is:

$$PF = \frac{P}{V_{RMS} \cdot I_{RMS}} = \frac{\int_0^T v(t) \cdot i(t) \cdot dt}{\sqrt{\int_0^T v(t)^2 dt} \cdot \sqrt{\int_0^T i(t)^2 dt}}$$

where T – period of $v(t)$.

If $v(t)$ is sinusoidal, real power is:

$$P = V_{RMS} \cdot I_{1RMS} \cdot \cos\varphi,$$

where I_{1RMS} - rms value of fundamental (first) harmonic of the current $i(t)$, φ - the phase angle (in radians) between $i_1(t)$ and $v(t)$.

Then power factor:

$$PF = \cos\varphi \cdot \frac{I_{1rms}}{I_{rms}}$$

The first term ($\cos\varphi$) represents phase displacement, the second one - current distortions. In linear circuits there are no higher harmonics: $I_{RMS} = I_{1RMS}$ and $PF = \cos\varphi$.

Total harmonic distortions (THD): the ratio between apparent power associated with higher order harmonics and fundamental harmonic is called. For sinusoidal voltage source:

$$\text{THD} = \frac{\sqrt{I_{RMS}^2 - I_{1RMS}^2}}{I_{1RMS}} = \sqrt{\frac{(\cos\varphi)^2}{PF^2} - 1}$$

2.2 CCM PFC BOOST

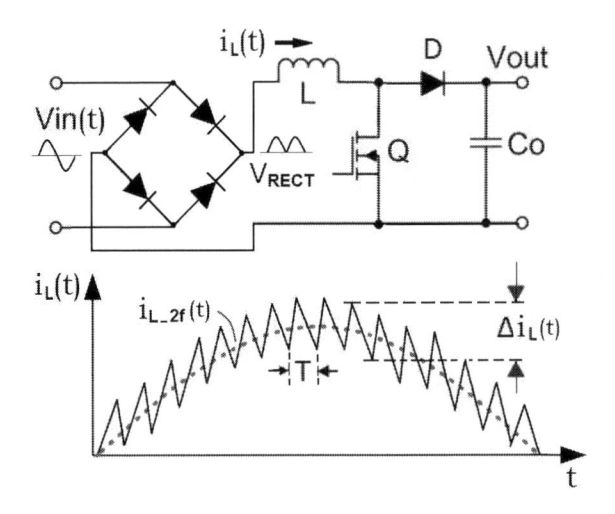

The power train of the circuit functions like in DC-DC boost converter (Chapter 1.4), except the input voltage V_{RECT} is rectified AC rather than DC. The power switch Q is driven ON and OFF by a control circuit at a frequency much higher than the input line frequency. The control circuit varies the duty cycle in such a way as to force twice the line frequency inductor current $i_{L\_2f}(t)$ to follow the shape of the input AC voltage while at the same time regulating output voltage **Vout**. In continuous conduction mode (CCM) the inductor current $i_L(t)$ does not fall to zero during a portion of AC cycle.

DUTY CYCLE

Neglecting rectifier bridge voltage drop, instantaneous value of the rectifier output voltage equals to absolute value of input AC voltage:

$$V_{RECT} \approx |Vin(t)| = \sqrt{2} \cdot V_{IN\_RMS} \cdot |\sin(2 \cdot \pi \cdot f_{IN} \cdot t)|,$$

where $|Vin(t)|$ – absolute value of input voltage; V_{IN_RMS} – rms value of input voltage, f_{IN} – input line frequency (in most applications f_{IN} is 50 or 60 Hz, in commercial aircrafts it is 360 to 800 Hz).

Neglecting the input voltage change during a switching pulse, rolling duty cycle varies over AC cycle as:

$$D(t) \approx 1 - \eta \cdot |Vin(t)|/Vout,$$

where η – converter efficiency; **Vout** - output voltage, which is always higher than amplitude of **Vin(t)**.

INDUCTOR L

From power balance, twice the line frequency f_{IN} component of inductor current, which is filtered $i_L(t)$ is given by:

$$i_{L\_2f}(t) = \frac{Pout}{\eta \cdot |Vin(t)| \cdot PF}$$

where Pout – output power, PF – power factor (PF<1).

Neglecting current ripple, RMS value of $i_{L\_2f}(t)$:
$I_{L\_RMS} = Pout/(\eta \cdot V_{IN\_RMS} \cdot PF)$, where V_{IN_RMS} – rms value of input voltage. Maximum I_{L_RMS} appears at low line.

DC component of inductor current, which determines DC copper losses: $I_{L\_DC} = I_{L\_RMS} \cdot 2\sqrt{2}/\pi$.

RMS value of AC component of inductor current, which determines low frequency ($2f_{IN}$) core and copper losses:

$$I_{L\_AC\_RMS} = \sqrt{I_{L\_RMS}^2 - I_{L\_DC}^2} = \frac{Pout}{\eta \cdot V_{IN\_RMS} \cdot PF}\sqrt{1 - \frac{8}{\pi^2}}$$

Rolling peak to peak ripple current in the inductor:

$$\Delta i_L(t) = \frac{|Vin(t)| \cdot D(t)}{L \cdot F} = \frac{|Vin(t)| \cdot (1 - \frac{\eta \cdot |Vin(t)|}{Vout})}{L \cdot F}$$

where F - switching frequency in hertz (F≫f_{IN}); L – inductance in henry.

Rolling RMS value of switching frequency component of inductor current: $\Delta i_{L\_RMS}(t) = \Delta i_L(t)/2\sqrt{3}$.
Averaging $\Delta i_{L\_RMS}(t)$ over AC line period yields:

$$\Delta i_{L\_RMS} = \frac{V_{IN\_RMS}}{L \cdot F} \cdot \sqrt{\frac{1}{12} + \frac{\eta^2 \cdot V_{IN\_RMS}^2}{8 \cdot Vout^2} - \frac{4\sqrt{2} \cdot \eta \cdot V_{IN\_RMS}}{9\pi \cdot Vout}}$$

This value determines core and copper coil losses at switching frequency F.

Inductor ripple current reaches maximum at the moment when the instantaneous input voltage $|Vin(t)|=Vout/2\eta$ (if input reaches this value):

$$\Delta i_{L\_MAX} = \frac{Vout}{4 \cdot L \cdot F \cdot \eta}$$

The condition of continuous conduction mode: $\Delta i_{L\_MAX}/2 < i_{L\_2f}$. In practice, to limit current ripple, L is selected such that at low line $\Delta i_{L\_MAX} \leq (0.3...0.6) \cdot i_{L\_PK}$, where i_{L_PK} is amplitude of the twice the line frequency current $i_{L\_2f}(t)$:

$$L \geq (1.5...3) \frac{\eta \cdot V_{IN\_MIN\_RMS}^2 \cdot \left(1 - \frac{\eta \cdot \sqrt{2} \cdot V_{IN\_MIN\_RMS}}{Vout}\right)}{F \cdot Pout}$$

Peak inductor current $i_{L\_2f}(t) + \Delta i_L(t)/2$ occurs at low line:

$$I_{L\_PK} = \frac{\sqrt{2} \cdot Pout}{\eta \cdot V_{IN\_MIN\_RMS} \cdot PF} + \frac{V_{IN\_MIN\_RMS}}{2 \cdot L \cdot F} \cdot D_{LL}$$

where D_{LL} – duty cycle at the peak of low line voltage:

$$D_{LL} = 1 - \frac{\eta \cdot \sqrt{2} \cdot V_{IN\_MIN\_RMS}}{Vout}$$

The inductor should provide the required inductance L at I_{L_PK}. See Chapter 4.4 for power inductor design.

SWITCH Q
Peak voltage: $V_{Q\_PK} = Vout$;
Peak current: $I_{Q\_PK} = I_{L\_PK}$;

Maximum RMS current (for simplicity neglecting ΔI_L and assuming ideal power factor correction PF=1.0):

$$I_{Q\_RMS} \approx \frac{P_{out}}{\eta \cdot V_{IN\_MIN\_RMS}} \sqrt{1 - \frac{1.2\eta \cdot V_{IN\_MIN\_RMS}}{V_{out}}}$$

RECTIFIER D
Peak voltage: $V_{D\_PK} = V_{out}$;
Average current: $I_{D\_AV} = P_{out}/V_{out}$;
Peak current: $I_{D\_PK} = \sqrt{2} \cdot P_{out}/(\eta \cdot V_{IN\_MIN\_RMS}) + \Delta I_L/2$;
Maximum RMS current (neglecting ΔI_L):

$$I_{D\_RMS} \approx P_{out} \cdot \sqrt{\frac{8\sqrt{2}}{3 \cdot \pi \cdot \eta \cdot V_{IN\_MIN\_RMS} \cdot V_{out}}}$$

CAPACITOR Co
Maximum net RMS current (neglecting ΔI_L):

$$I_{Co\_RMS} = \frac{P_{out}}{V_{out}} \cdot \sqrt{\frac{8\sqrt{2} \cdot V_{out}}{3\pi \cdot \eta \cdot V_{IN\_MIN\_RMS}} - 1}$$

Maximum RMS value of switching frequency component:

$$I_{Co\_SW\_RMS} = \frac{P_{out}}{V_{out}} \cdot \sqrt{\frac{8\sqrt{2} \cdot V_{out}}{3\pi \cdot \eta \cdot V_{IN\_MIN\_RMS}} - 1.5}$$

RMS value of twice the line frequency current:

$$I_{Co\_2f\_RMS} = P_{out}/(\sqrt{2} \cdot V_{out})$$

During AC power interruption the capacitance Co supplies the load. Its value is selected to maintain the PFC output voltage above minimum acceptable value **Vout_min** during required hold-up time T_{HOLD}:

$$Co > 2 \cdot P_{out} \cdot T_{HOLD}/(V_{out}^2 - V_{out\_min}^2),$$

where **Vout_min** is selected above dropout voltage of downstream converter; T$_{HOLD}$ is typically selected >20ms for mains applications, or >400ms for commercial aircrafts.

INPUT CAPACITOR

The low frequency and high frequency of the current in input capacitor (not shown on the diagram) are $I_{L\_AC\_RMS}$ and Δi_{L_RMS} respectively. The net RMS current:

$$I_{Cin\_RMS} = \sqrt{I_{L\_AC\_RMS}^2 + \Delta i_{L\_RMS}^2}$$

2.3 DCM PFC BOOST

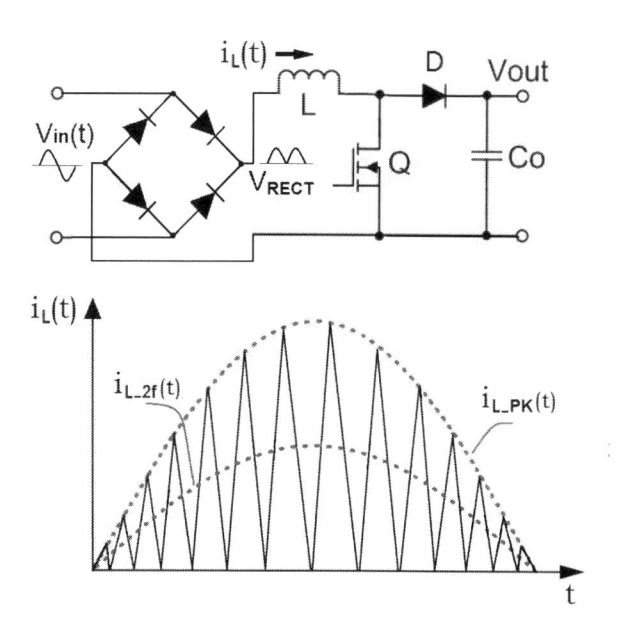

The power train of the circuit functions similarly to that of CCM boost converter, except inductor current i_L always falls to zero before the next switching cycle. In a typical **critical** (a.k.a **boundary**) **conduction mode** control implementation, each switching cycle begins when i_L reaches to zero, and the switching frequency varies with line and load conditions. A control circuit forces the envelope of inductor peak currents $i_{L\_PK}(t)$ to follow rectified input voltage waveform $Vin(t)$ while regulating output voltage $Vout$. Such operation features lower switching losses. Power switch Q turns on at zero current, and when instantaneous input voltage is below half of $Vout$, it can turn on at zero voltage.

SWITCHING FREQUENCY
Neglecting rectifier bridge voltage drop, instantaneous value of the rectifier voltage equals to absolute value of input AC voltage: $V_{RECT} \approx |Vin(t)|$.
Rolling on-time of each pulse: $ton(t) = L \cdot i_{L\_PK}(t)/|Vin(t)|$, where L – inductance in henry, $i_{L\_PK}(t)$ – rolling peak current of the inductor.

Rolling off-time of each pulse: toff(t)= L·i$_{L\_PK}$(t)/(Vout-|Vin(t)|).

Envelope of peak currents: i$_{L\_PK}$(t)=2·i$_{L\_2f}$(t), where i$_{L\_2f}$(t) – twice the line frequency inductor current (which is moving midpoint of the current ramps). From power balance:

$$i_{L\_2f}(t) = \frac{\sqrt{2} \cdot \text{Pout}}{\eta \cdot V_{IN\_RMS} \cdot PF} \cdot |\sin(2\pi \cdot f_{IN} \cdot t)|$$

where V$_{IN\_RMS}$ – rms value of input voltage, f$_{IN}$ – input line frequency (in most applications f$_{IN}$ is 50 or 60 Hz, in commercial aircrafts it is 360 to 800 Hz); η –converter efficiency; Pout – output power; PF – power factor (PF<1).

At a given power and input voltage, on-time is constant:

$$ton = \frac{2 \cdot L \cdot \text{Pout}}{\eta \cdot V_{IN\_RMS}^2 \cdot PF}$$

OFF-time is modulated over AC cycle:

$$toff(t) = \frac{2\sqrt{2} \cdot L \cdot \text{Pout} \cdot |\sin(2\pi \cdot f_{IN} \cdot t)|}{Vout - \sqrt{2} \cdot V_{IN\_RMS} \cdot |\sin(2\pi \cdot f_{IN} \cdot t)|}$$

Assuming for simplicity unity power factor, period of switching cycles varies according to:

$$T(t) = \frac{2 \cdot L \cdot \text{Pout} \cdot Vout}{\eta \cdot V_{IN\_RMS}^2 \cdot (Vout - \sqrt{2} \cdot V_{IN\_RMS} \cdot |\sin(2\pi \cdot f_{IN} \cdot t)|)}$$

Switching frequency then varies according to:

$$F(t) = \frac{\eta \cdot V_{IN\_RMS}^2 \cdot (Vout - \sqrt{2} \cdot V_{IN\_RMS} \cdot |\sin(2\pi \cdot f_{IN} \cdot t)|)}{2 \cdot L \cdot \text{Pout} \cdot Vout}$$

Switching frequency reaches peak at high line near zero crossing:

$$F_{MAX} = \frac{\eta \cdot V_{IN\_MAX\_RMS}^2}{2 \cdot L \cdot Pout}$$

INDUCTOR L

For selected maximum frequency **Fmax**, required inductance:

$$L > \frac{\eta \cdot V_{IN\_MAX\_RMS}^2}{2 \cdot Fmax \cdot Pout}$$

Peak current at low line:

$$I_{L\_PK} = \frac{2\sqrt{2} \cdot Po}{\eta \cdot V_{IN\_MIN\_RMS} \cdot PF}$$

RMS current as low line:

$$I_{L\_RMS} = \frac{Po}{\sqrt{3} \cdot \eta \cdot V_{IN\_MIN\_RMS} \cdot PF}$$

DC component of inductor coil current, which determines DC copper losses:

$$I_{L\_DC} = \frac{\sqrt{2} \cdot Po}{\pi \cdot \eta \cdot V_{IN\_MIN\_RMS}}$$

RMS value of AC component of inductor current, which determines core losses and AC copper coil losses:

$$I_{L\_AC\_RMS} = \frac{Po}{\eta \cdot V_{IN\_MIN\_RMS} \cdot PF} \cdot \sqrt{\frac{1}{3} - \frac{2}{\pi^2}}$$

The inductor has to provide required L at I_{L_PK}. See Chapter 4.4 for power inductor design.

For simplicity, in the following equations we assume unity power factor.

RECTIFIER D
Peak voltage: $V_{D\_PK}=V_{out}$;
Average current: $I_{D\_AV}=P_{out}/V_{out}$;
Peak current: $I_{D\_PK}=I_{L\_PK}$;
Maximum RMS current:

$$I_{D\_RMS} = \frac{1.27 \cdot P_o}{\sqrt{\eta \cdot V_{IN\_MIN\_RMS} \cdot V_{out}}}$$

CAPACITOR Co
Maximum RMS current:

$$I_{Co\_RMS} = I_{out} \cdot \sqrt{\frac{1.27 \cdot V_{out}}{\eta \cdot V_{IN\_MIN\_RMS}} - 1}$$

During AC power interruption the capacitance Co supplies the load. Its value is selected to maintain the PFC output voltage above minimum acceptable value **Vout_min** during required hold-up time T_{HOLD}:

$$Co > 2 \cdot P_{out} \cdot T_{HOLD}/(V_{out}^2 - V_{OUT\_MIN}^2),$$

where V_{OUT_MIN} is selected above dropout voltage of downstream converter.
The required T_{HOLD} is typically selected ≥ 20ms for mains applications, or ≥ 400ms for commercial aircrafts.

3 FEEDBACK LOOP FUNDAMENTALS

3.1 BODE PLOT AND STABILITY CRITERIA

In a generic single-loop system with negative feedback Vout=G·(Vref-Vout·H), where G is a transfer function of the system and H is the feedback circuit.

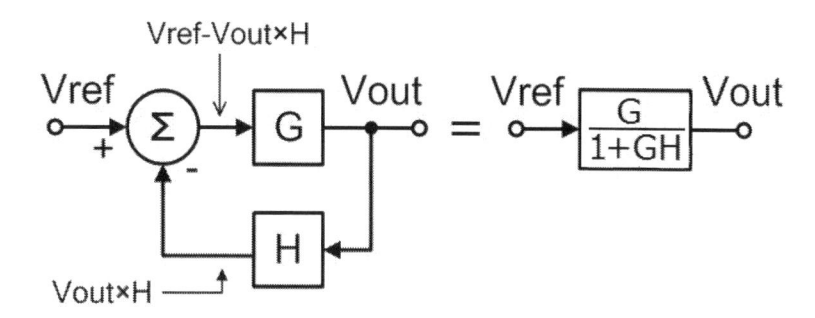

Solving this for **Vout** gives closed loop transfer function from reference to output:

$$\frac{Vout}{Vref} = \frac{G}{1+GH} = \frac{1}{H} \cdot \frac{T}{1+T}$$

where T=G·H - open loop gain.
Frequency characteristic of open loop gain T(s) determines stability of a closed-loop system.

At low frequencies when $|T|>>1$, and the closed loop gain Vout/Vref≈1/H is independent of the gain G in the forward path of the loop. **Vout** is stable and follows reference via feedback ratio H.

At high frequencies $|T|<<1$, Vout/Vref≈G and the feedback has no effect on the system.

When **T=-1**, i.e. its magnitude is one and phase shift is -180⁰ in addition to -180⁰ due to negative feedback, the closed loop gain mathematically approaches infinity (in reality, due to the nonlinearities, the gain will have a high but finite value). When **T=-1** the power supply will oscillate because any disturbance injected into the loop will propagate around the feedback loop and return in phase with, and equal in amplitude to the original injected signal.

For convenience, the magnitude is plotted logarithmically: $T_{dB}=20\log(T)$.

The graph of the logarithmic magnitude gain and phase of an open loop system vs. logarithmic frequency is referred to as **Bode plot**.

Bode plot of a typical power supply may look like this:

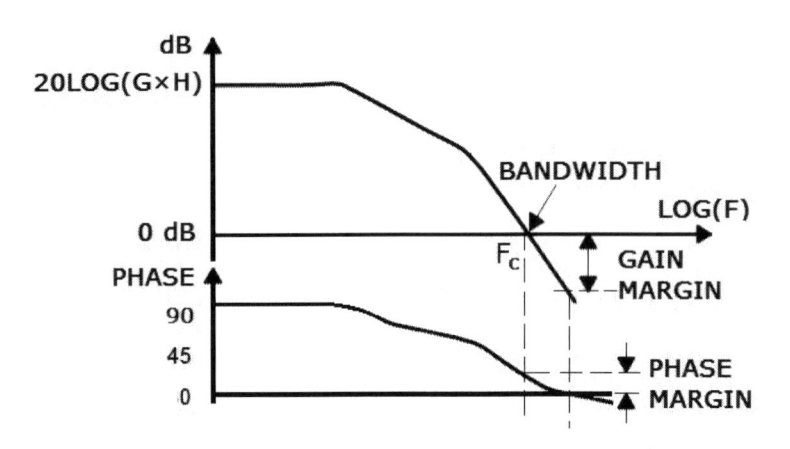

Nyquist Stability Criterion adapted to Bode plot: A system with negative feedback is stable if at zero dB crossing of the open loop magnitude the phase is not be equal to +/-180⁰ (besides 180° phase shift from negative feedback).

Phase and gain margins of open loop describe safety margin of stability of the power supply and how it responds to perturbations.

The **phase margin** is the amount of phase at the open-loop crossover frequency Fc (which is also called bandwidth). At crossover frequency closed-loop gain is -3 dB.

The best compromise between loop stability and response time is achieved with the phase margin between 45^0 and 75^0.

The **gain margin** is magnitude at zero phase (which is -180^0 phase shift in addition to -180^0 phase shift at the origin due to negative feedback). A typical gain margin selection is 10 dB, although anything between 3 dB and 12 dB is usually acceptable.

3.2 LAPLACE TRANSFORM: ZEROES AND POLES

Laplace transform allows converting linear differential equations in time domain into algebraic expressions. The solution can then be transformed back to the time domain with the inverse Laplace transform. By definition, unilateral Laplace transform of an arbitrary function $x(t)$ is:

$$G(s) = L\{(x(t)\} = \int_0^\infty x(t)e^{-st}dt$$

where $s=\sigma+j\omega$ is so-called complex frequency, $\omega=2\pi F$ - the angular frequency in radians per seconds, F – frequency in hertz, j - imaginary unit. For steady state analysis we put $\sigma=0$ and $s=j\omega$.

A typical small signal transfer function of a linear system can be presented in the factored form as:

$$G(s) = \frac{(1+s/z_1)(1+s/z_2) \ldots (1+s/z_M)}{s/p_0 \cdot (1+s/p_1)(1+s/p_2) \ldots (1+s/p_N)}$$

The above function have **zeroes** (numerator roots) at $s=-z_1, -z_2,\ldots -z_M$ and poles (denominator roots) at $s=-p_1, -p_2,\ldots -p_N$, as well as a pole p_0 in the origin. All poles and zeros are real numbers.

SINGLE POLE
The function: $G_{1P}(s)=1/(1+s/p)$, where p - pole frequency;
The logarithmic gain: $G_{1P\_dB}(\omega)=-10\cdot\log(1+(\omega/p)^2)$;
The phase shift: $\arg(G_{1P}(s))=-\arg(1+s/p)=-\arctan(\omega/p)$.

At $\omega<<p$: $G_{1P\_dB}=0$,
at $\omega=p$ (cutoff frequency): $G_{1P\_dB}=-3dB$; the phase shift is $-45°$;
at $\omega>>p$: $G_{1P\_dB}=-20\cdot\log(\omega/p)$, gain G_{1P_dB} is falling with 20 dB/decade slope (-1 slope). When $\omega\to\infty$, the phase shift is $-90°$.

TWO-POLE

The function:

$$G_{2P}(s) = \frac{1}{\left(1+\frac{s}{p_1}\right)\left(1+\frac{s}{p_2}\right)} = \frac{1}{1+\frac{s}{\omega_R \cdot Q}+\left(\frac{s}{\omega_R}\right)^2}$$

where $\omega_R=\sqrt{(p_1 \cdot p_2)}$ - resonant (cutoff) frequency; $Q = \frac{\sqrt{p_1 \cdot p_2}}{p_1+p_2}$ - quality factor.

Physical meaning of quality factor: a measure of the losses in the system[2]:

$$Q = 2\pi \cdot \frac{\text{Peak stored energy}}{\text{Peak disipated energy}}$$

The logarithmic gain:

$$G_{2P\_dB}(\omega) = -10\log\left[\left(1-\frac{\omega^2}{\omega_R^2}\right)^2 + \left(\frac{\omega}{\omega_R \cdot Q}\right)^2\right]$$

The phase:

$$\arg(G_{2P}(s)) = -\arctan\left[\frac{\frac{\omega}{\omega_R \cdot Q}}{1-\left(\frac{\omega}{\omega_R}\right)^2}\right]$$

At $\omega \ll \omega_R$: $G_{2P\_dB}=0$,
at $\omega = \omega_R$: $G_{2P\_dB}=20\log(Q)$; the phase shift is $-90°$;
at $\omega \gg \omega_R$: $G_{2P\_dB} \approx -40 \cdot \log(\omega/\omega_R)$, gain G_{2P_dB} is falling with 40 dB/decade slope (-2 slope). When $\omega \to \infty$, the phase shift is $-180°$.

To be stable any closed loop system should have open-loop gain crossing zero dB at a slope less than -40 dB/decade to keep phase lag $<180°$. **For optimum margins, the slope near zero dB crossing should tend to be around -20 dB/decade.**

ZERO

The function: $G_{1Z}(s)=1+s/z$, where z - zero frequency;
The logarithmic gain: $G_{1Z\_dB}(\omega)=10 \cdot \log(1+(\omega/z)^2)$;
The argument: $\arg(G_{1Z}(s))=\arg(1+s/z)=\arctan(\omega/z)$.

At $\omega \ll z$: $G_{1Z\_dB}=0$;
at $\omega=z$ (cutoff frequency): $G_{1Z\_dB}=+3dB$; the phase boost is $+45°$;

at $\omega\gg z$: $G_{1Z\_dB}=20\cdot\log(\omega/z)$, gain G_{1Z_dB} is increasing with 20 dB/decade slope (+1 slope); when $\omega\to\infty$, the phase shift is +90°.

Note. Poles and zeros at the origin introduce a fixed phase shift of -90° and +90° each respectively.

RIGHT–HALF–PLANE ZERO

The function: $G_{RHPZ}(s)=1-s/z$, where z – right-half-plane zero frequency;
The logarithmic gain: $G_{RHPZ\_dB}(\omega)=10\cdot\log(1+(\omega/z)^2)$;
The argument: $\arg(G_{RHPZ}(s))=-\arg(1+s/z)=-\arctan(\omega/z)$.

At $\omega\ll z$: $G_{RHPZ\_dB}=0$;
at $\omega=z$ (cutoff frequency): $G_{Z\_dB}=+3dB$; the phase shift is -45°;
at $\omega\gg p$: $G_{RHPZ\_dB}=20\cdot\log(\omega/z)$, gain G_{RHPZ_dB} is increasing with 20 dB/decade slope (+1 slope); when $\omega\to\infty$, the phase shift is -90°.

3.3 TRANSIENT RESPONSE vs. PHASE MARGIN

An open-loop transfer function of a stabilized converter (made of the power stage and a compensator) near crossover frequency Fc can usually be approximated by second-order function with one pole at the origin ω_0 and one high-frequency pole ω_1 above crossover frequency:

$$T_{OL}(s) = \frac{1}{(s/\omega_0)(1 + s/\omega_1)}$$

Substituting $s=j\omega$ and solving $|T_{OL}(j\omega_c)|=1$, we obtain crossover frequency of such system:

$$\omega_c = \frac{\omega_1 \cdot \sqrt{\sqrt{1 + 4 \cdot (\omega_0/\omega_1)^2} - 1}}{\sqrt{2}}$$

Phase margin:

$$\varphi_M = \arctan \frac{\omega_1}{\omega_c} = \arctan \left(\frac{\sqrt{2}}{\sqrt{\sqrt{1 + 4 \cdot (\omega_0/\omega_1)^2} - 1}} \right)$$

Quality factor relates to phase margin φ_M [3]:

$$Q = \frac{\sqrt{\cos\varphi_M}}{\sin\varphi_M}$$

The closed-loop of the above second-order transfer function (assuming for simplicity feedback divider H=1):

$$T_{CL}(s) = \frac{T_{OL}(s)}{1 + T_{OL}(s)} = \frac{1}{1 + s/Q \cdot \omega_R + s^2/\omega_R^2}$$

where $\omega_R = \sqrt{(\omega_0 \cdot \omega_1)}$ – resonant (cutoff) frequency; $Q = \sqrt{(\omega_0/\omega_1)}$ - quality factor.

A transient response of a closed-loop system (its response to unit step function) is inverse Laplace transform of the product of $T_{CL}(s)$ by $1/s$. A general result for 2nd order function is calculated in [4]. Here is an example of the response to reference voltage unit step for different phase margins φM for crossover frequency $Fc=10$ kHz.

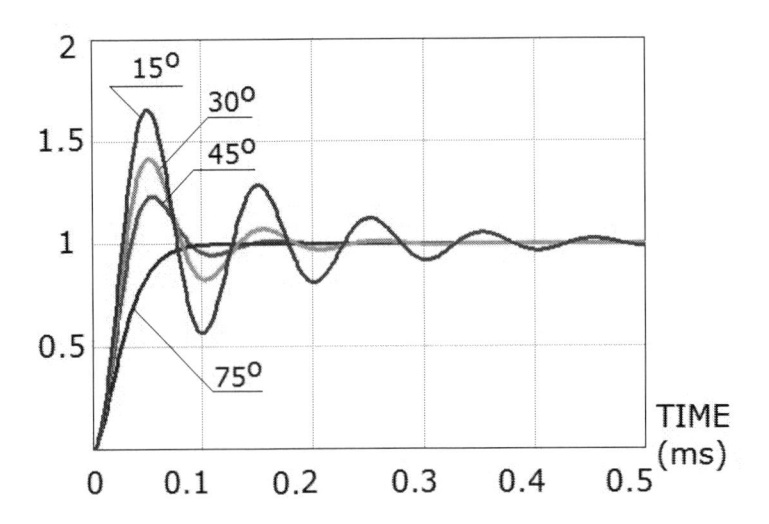

Phase margin near 75° produces critically damped response. A good compromise between response time and stability margin with small overshoot is achieved around 45°. In real systems overshoot may be lower because of limited slew rate of the stimulus. But qualitatively, the response curves for different phase margins will be similar.

3.4 CHOOSING CROSSOVER FREQUENCY

Crossover frequency **Fc** (open loop zero db crossing) is selected primarily based on maximum output deviation under dynamic loading and required response time. It also depends on switching frequency, output filter resonant frequency, and right-half plane zero (RHPZ) frequency.

Voltage deviation ΔVout at step load consists of the capacitance and ESR terms. Maximum deviation generally depends on the relationship between bandwidth and filter parameters and on type of the feedback control. In practical designs with large output capacitor and low-ESR ceramic capacitors in parallel, the response is determined mainly by charge/discharge of output capacitors. In such systems the peak deviation is reached at some time after the load step and is a function of the control method. The following equations are given for small-signal conditions (i.e. for the transients that do not saturate error amplifier) [5].

VOLTAGE MODE CONTROL

Peak transient deviation:

$$\Delta V_{OUT} \approx \frac{1 + (4 \cdot Co \cdot ESR \cdot Fc)^2}{8 \cdot Co \cdot Fc} \cdot \Delta I_{OUT}$$

With ceramic output capacitors net ESR is very low and $\Delta V_{out} \approx \Delta I_{OUT}/(8 \cdot Co \cdot Fc)$

Required crossover frequency for desired maximum ΔVout:
$$Fc \geq \Delta I_{OUT}/(8 \cdot Co \cdot \Delta V_{out})$$

CURRENT MODE CONTROL

Peak transient deviation ΔVout:

$$\Delta V_{OUT} \approx \Delta I_{OUT}/(2\pi \cdot Co \cdot Fc)$$

The required crossover frequency for desired maximum ΔVout:

$$Fc \geq \Delta I_{OUT}/(2\pi \cdot Co \cdot \Delta Vout)$$

The response time (time to peak deviation after which the output begins recovering):
$$T_R \approx 1/(4 \cdot Fc)$$

Selection of **Fc** to meet desired response time:
$$Fc \geq 1/(4 \cdot T_R)$$

Crossover frequency **Fc** should be kept below at least 1/2 of switching frequency (we usually set **Fc≤F/5...10**), and below 1/3 of right half plane zero (if present). For voltage mode control of buck-derived topologies it should also stay ≥3...5 times above resonance of output LC filter.

3.5 LOOP COMPENSATION BASICS

The complete voltage mode control loop may contain several cascaded stages in the forward path:

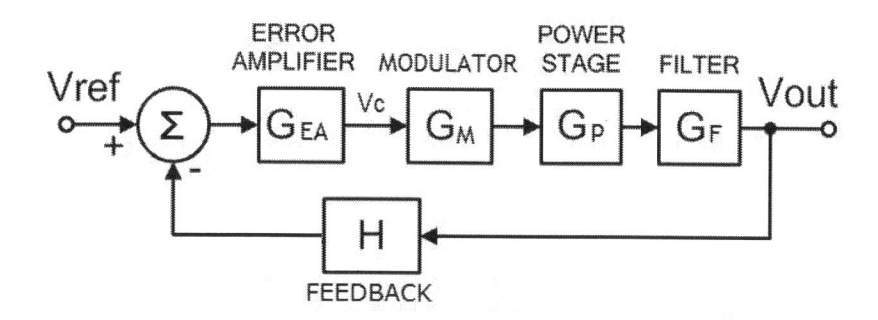

The logarithmic open loop gain in decibel:
$$T_{dB} = G_{EA\_dB} + G_{M\_dB} + G_{P\_dB} + G3_{F\_dB} + H_{dB}$$

The net open loop phase shift:
$$\varphi = \varphi_{EA} + \varphi_M + \varphi_P + \varphi_F + \varphi_H$$

COMPENSATION DESIGN BASICS

A typical design goal is to achieve desired open loop crossover frequency with phase margin between 45° and 75°, gain margin 6-12 dB, and with high gain at low frequencies.
The basic steps are:
1. Choose crossover frequency **Fc** (see Chapter 3.4);
2. Derive transfer functions of modulator, power stage and filter (G_M, G_P, G_F), and find net differential transfer function from control signal **Vc** to output: $d(Vout)/dVc$;
3. Find net open loop gain and phase of the control to output transfer function at **Fc** from the equation in step 2 or by plotting net logarithmic gain and phase (a piecewise linear approximation can be used when sketching a Bode plot by hand);
4. Once **Vc** to **Vout** gain and phase at **Fc** are known, construct error amplifier compensation scheme (compensator) such that at **Fc** the net open loop gain is equal to 1 (0dB), phase

margin is between 45° and 75°, and the gain is high in DC for low static error. For AC-powered applications, the gain should remain high at twice the line frequency for low output ripple. Since different converters and different control methods yield different transfer functions, unfortunately, there is no single unified procedure for the compensator design.

3.6 VOLTAGE MODE CONTROL TRANSFER FUNCTIONS

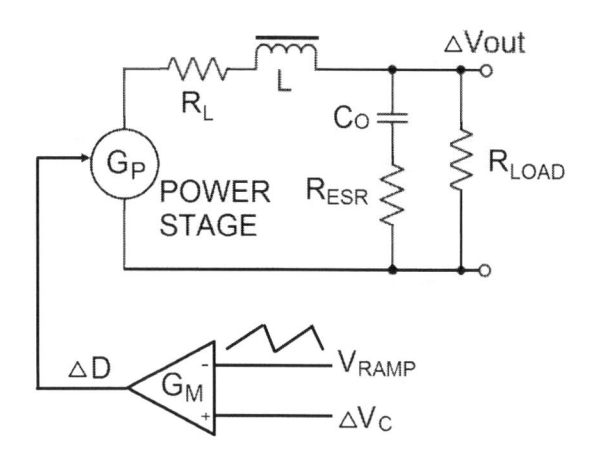

MODULATOR

Modulator varies a parameter of switching pulses (duty cycle D for fixed frequency or frequency F for variable frequency) by comparing control voltage Vc from the error amplifier to a fixed ramp. Modulator's small signal gain is change of the controlled parameter vs. change in control voltage:

$$G_M = \Delta D/\Delta Vc \text{ or } G_M = \Delta F/\Delta Vc$$

For fixed-frequency duty cycle modulator:

$$G_M = D_{MAX}/V_{RAMP\_PK}$$

where D_{MAX} – maximum duty cycle (between 0.45 and 0.99 for most controllers), V_{RAMP_PK} - peak-to-peak amplitude of the V_{RAMP} sawtooth.

For frequency modulator (like in LLC) assuming frequency varies linearly with control voltage:

$$G_M = \Delta F_{MAX}/V_{RAMP\_PK},$$

where ΔF_{MAX} – maximum range of frequency variation.

POWER STAGE

With duty cycle control, power stage DC gain from duty cycle D to output is given by differentiating:

$$G_P = dVout(D)/dD$$

If output voltage is a function of frequency F, power stage gain

$$G_P = dVout(F)/dF$$

See Chapter 1 for transfer functions **Vout(D)** and **Vout(F)** for various topologies.

OUTPUT LC FILTER

The filter transfer function is determined by LC and load resistance:

$$G_F(s) = \frac{1 + sR_{ESR}C_o}{1 + s\left(\frac{L}{R_{LOAD}} + \frac{R_L}{R_{LOAD}}R_{ESR}C_o + R_{ESR}C_o\right) + s^2\left(\frac{R_{ESR}C_oL}{R_{LOAD}} + C_oL\right)}$$

If $L/R_{LOAD} \gg C_o \cdot R_{ESR}$ and $R_{LOAD} \gg R_{ESR}$:

$$G_F(s) \approx \frac{1 + \frac{s}{\omega_Z}}{1 + \frac{s}{\omega_0 \cdot Q} + \frac{s^2}{\omega_0^2}}$$

where:

$\omega_0 = 1/\sqrt{L \cdot C_o}$ - filter resonant frequency;

$Q = \dfrac{R_{LOAD}}{\sqrt{L/C_o}}$ - quality factor;

$\omega_Z = 1/(R_{ESR} \cdot C_o)$ – zero associated with ESR.

3.7 CURRENT MODE CONTROL TRANSFER FUNCTIONS

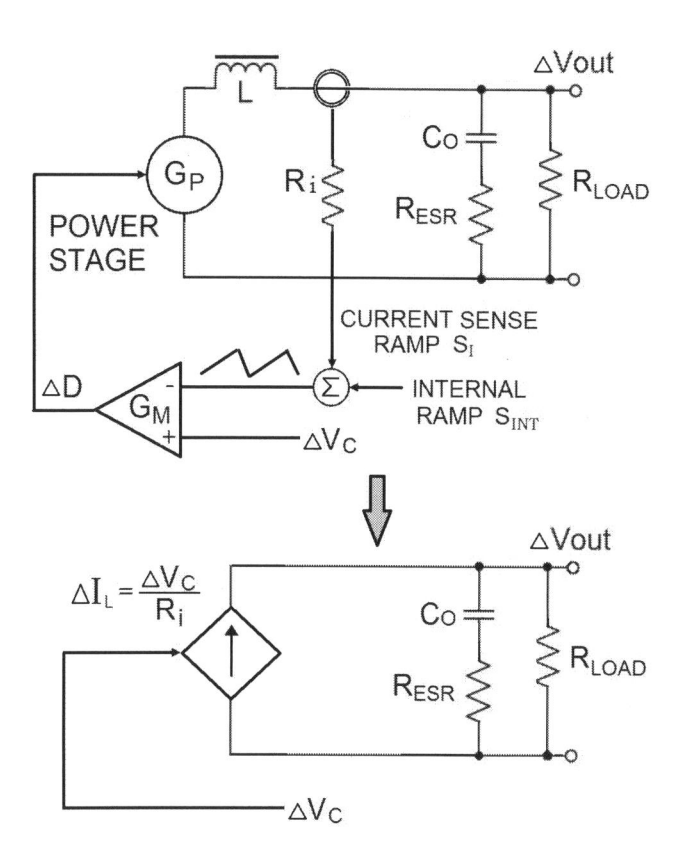

POWER STAGE

The outer open loop DC gain G_P is the same as for voltage-mode control. With the inner current feedback:

$$\Delta V_{out} = (\Delta V_C - \Delta I_L \cdot R_i) \cdot G_M \cdot G_P.$$

Substituting for $\Delta I_L = \Delta V_{out}/R_{LOAD}$ and solving for ΔV_{out} gives small signal DC gain from control to output:

83

$$\frac{\Delta V_{out}}{\Delta V_c} = \frac{G_M \cdot G_P \cdot R_{LOAD}}{R_{LOAD} + R_i \cdot G_M \cdot G_P}$$

If $R_i \cdot G_M \cdot G_P \gg R_{LOAD}$, small-signal DC gain:

$$\Delta V_{out}/\Delta V_c \approx R_{LOAD}/R_i.$$

At this, the inductor is effectively removed from the outer voltage control loop and acts as a voltage-controlled current source with gain $1/R_i$.

MODULATOR

In peak current mode control (CMC) the ramp is provided by sensed inductor current I_L. In continuous conduction operation with duty cycle approaching or exceeding 0.5, a portion of internal ramp should be added to I_L for loop stability (slope compensation). The error amplifier output (V_c) determines the peak level of the combined ramp, and the modulator accordingly varies a parameter of switching pulses. For duty cycle constant-frequency control, small signal gain of the CMC modulator [6]:

$$G_M = \frac{\Delta D}{\Delta V_c} = \frac{1}{(S_{I\_UP} + S_{INT}) \cdot T}$$

where S_I – up-slope of the current sense voltage at the input of PWM comparator (volt/sec); S_{INT} – up-slope of internal ramp (volt/sec); $T = 1/F$ – period of the switching frequency.

The up-slope of current sense voltage:

$$S_{I\_UP} = \frac{\Delta I_L}{t_{ON}} \cdot R_i = \frac{V_{L\_ON}}{L} \cdot R_i$$

where V_{L_ON} – voltage across inductor L during on-time; R_i – equivalent sense resistance (ratio between the voltage at the input of modulator and sensed inductor current, volts/amps).
If inductor current is sensed directly across sense resistor R_S, like in non-isolated convertors, $R_i = G_S \cdot R_S$, where R_S – current sense resistor, G_S – voltage gain from R_S to modulator input. In transformer isolated topologies secondary side inductor current is often sensed indirectly on primary side. In general, $R_i = G_S \cdot G_L \cdot R_S$, where G_L – ratio between current across R_S to inductor current.

If Rs is placed in series with primary side switch: $G_L = N2/N1$, where N1, N2 – power transformer primary and secondary turns respectively. Primary current sometimes is sensed via a current sense transformer with a single turn primary and some secondary turns Ncs. If Rs is placed across the output of such current sense transformer, the $G_L = N2/(N1 \cdot Ncs)$.

SLOPE COMPENSATION

Compensation ramp can be produced from PWM ramp pin (C_T) if the controller has it, or from gate drive output:

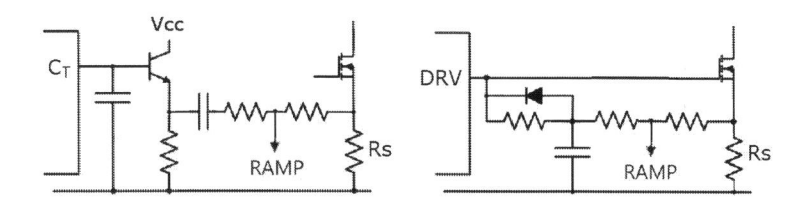

Optimal slope compensation ramp S_{INT} is a function of absolute value of the current sense voltage down-slope, which is given by:

$$S_{I\_DOWN} = \frac{|\Delta I_L|}{t_{OFF}} \cdot Ri \approx \frac{Vout}{L} \cdot Ri$$

For current loop stability we may select $S_{INT} = 0.5...0.8 \cdot S_{I\_DOWN}$.

Note that some PWM controllers provide slope compensation internally. Also, in transformer isolated topologies, magnetizing current of the power transformer automatically adds certain slope to the output inductor current reflected to the primary. Depending on magnetizing inductance and D_{MAX}, this may already be sufficient slope compensation.

FILTER

With inductor as a current source, the filter AC transfer function is determined by output capacitor and load resistance:

$$G_F(s) = \frac{\Delta Vo}{\Delta I_L} = \frac{1 + \frac{s}{\omega_Z}}{1 + s \cdot (\frac{1}{\omega_P} + \frac{1}{\omega_Z})}$$

where $\omega_Z = 1/(R_{ESR} \cdot C_O)$ – zero associated with ESR;
$\omega_P = 1/(R_{LOAD} \cdot C_O)$ – dominant pole formed by output capacitance and the load resistance.
If $\omega_Z \gg \omega_P$: $G_F(s) \approx R_{LOAD}/(1+s/\omega_P)$.

The net small signal gain from control to output voltage:

$$\frac{\Delta V_O}{\Delta V_C}(s) \approx \frac{R_{LOAD}}{R_i} \cdot \frac{1}{1 + s \cdot R_{LOAD} \cdot C_O}$$

An advanced model proposed in [6] has an additional "sampling gain" double-pole at half of switching frequency $F/2$.

3.8 INCREASING PHASE MARGIN

If phase margin is not sufficient, a phase boost at crossover frequency **Fc** can be created by a pair of zero and pole. A pole lags the phase by -90° and a zero boosts the phase by +90°. A pole-zero pair can change the phase at a certain frequency from 0° if the pole and the zero coincide to 90° if they are far apart.

Transfer function of zero-pole pair in general:

$$G_{Z-P}(s) = \frac{1 + s/s_Z}{1 + s/s_P}$$

where $s_Z = j \cdot 2\pi F_Z$ - zero, $s_P = j \cdot 2\pi F_P$ - pole.

Phase shift of such a pair at a given frequency F:

$$\Delta\varphi(F) = \arctan\left(\frac{F}{F_Z}\right) - \arctan\left(\frac{F}{F_P}\right)$$

For phase boost we select **Fz<Fc and Fp>Fc**. Peak of phase shift is achieved at $F_{PK} = \sqrt{(F_Z \cdot F_P)}$. To get a desired phase boost $\Delta\varphi$ at **Fc**, the pole and zero should be chosen as follows ([4]):

$$F_Z = F_C \cdot \sqrt{\frac{1 - \sin(\Delta\varphi)}{1 + \sin(\Delta\varphi)}}$$

$$F_P = F_C \cdot \sqrt{\frac{1 + \sin(\Delta\varphi)}{1 - \sin(\Delta\varphi)}}$$

The effect of phase margin on transient response is analyzed in [7].

3.9 ERROR AMPLIFIER GAIN WITH TL431

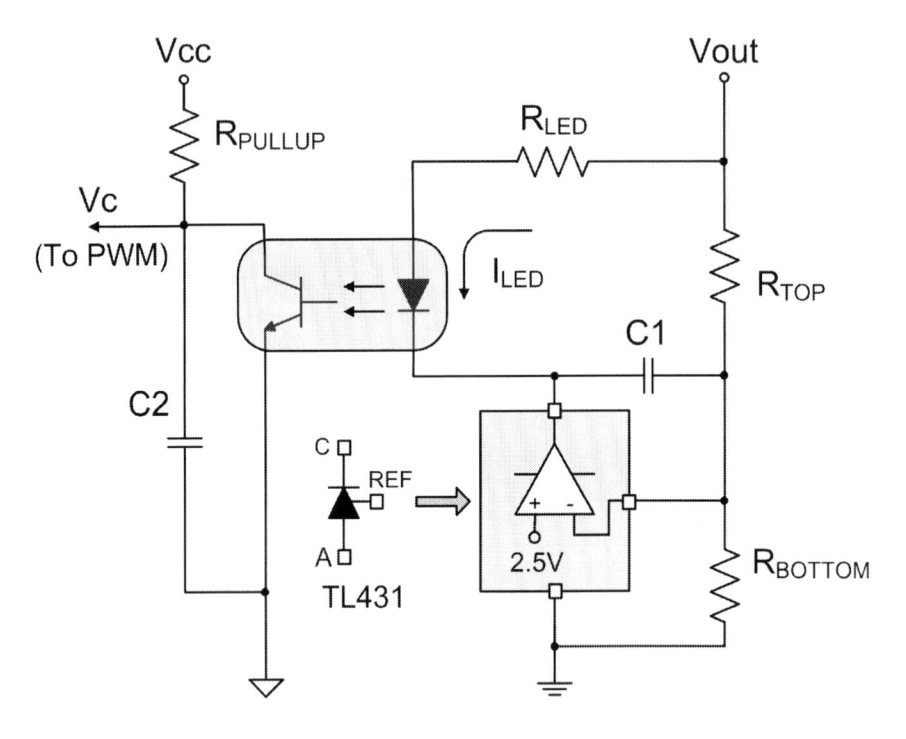

The three-terminal shunt regulator TL431 is often used in a feedback compensation network for isolated power converters. It contains an open-collector operational amplifier and 2.5V internal reference (a similar TLV431 has 1.24V reference). The internal circuitry is biased from the cathode. When the voltage on the REF pin exceeds the internal reference, TL431 sinks current from its cathode. Small-signal TL431 voltage response:

$$\text{Vcathode}(s) = -\text{Vout}(s) \cdot \frac{1}{s \cdot R_{TOP} C1}$$

Neglecting the LED impedance, its current:

$$I_{LED}(s) = \frac{\text{Vout}(s) - \text{Vcathode}(s)}{R_{LED}} = \frac{\text{Vout}(s)}{R_{LED}} \cdot \frac{1 + s \cdot R_{TOP} C1}{s \cdot R_{TOP} C1}$$

Small-signal control error voltage:

$$Vc(s) = \frac{I_{LED}(s) \cdot CTR}{1 + s \cdot R_{PULLUP}C_{POLE}}$$

where **CTR** - current transfer ratio of the optocoupler; $C_{POLE}=C2+C_{OPTO}$; C_{OPTO} – internal capacitance of the optocoupler. Some datasheet provide C_{OPTO}. Other may provide optocoupler's fall time at a certain load R_L. Its capacitance can be estimated as: $C_{OPTO}=Tf/(2.2 \cdot R_L)$. Generally, C_{OPTO} varies with current.

Output to control small-signal gain $G_{EA}(s)=Vc(s)/Vout(s)$:

$$G_{EA}(s) = -\frac{R_{PULLUP} \cdot CTR}{R_{LED}} \cdot \frac{1 + s \cdot R_{TOP}C1}{s \cdot R_{TOP}C1 \cdot (1 + s \cdot R_{PULLUP}C_{POLE})}$$

4 MAGNETICS DESIGN

4.1 TRANSFORMER CALCULATIONS

For rectangular pulses from Faraday's law:
$$Vin = \frac{\Delta B \cdot N1 \cdot Ac \cdot F \cdot 10^{-8}}{D}$$

where Vin – voltage across transformer's primary, **N1**- primary turns, **Ac**- core's equivalent cross-sectional area in sq.cm, D - duty cycle, F – switching frequency in hertz, **ΔB** – flux swing (gauss).
Net current density (amps/sq.cm) in primary and secondary coils is:

$$J = 2 \cdot N1 \cdot I_{1RMS}/(Aw \cdot K)$$

where I_{1RMS} - rms value of primary current, **Aw** - window area available for the windings (sq.cm); **K** - winding fill factor.
Eliminating **N1** from the above equations yields required product of magnetic cross-section area **Ac** by the window area **Aw**:

$$Ap = Ac \cdot Aw = \frac{2 \cdot D \cdot Pin \cdot 10^8}{K \cdot J \cdot F \cdot \Delta B}$$

where **Ap** - area product in cm^4, Pin= $Vin \cdot I_{1RMS}$ - transformer's input power.

Core size is selected based on required **Ap** (magnetic areas **Ac** are provided in the cores datasheets, **Aw** values can be found in the matching bobbins datasheets).
Since duty cycle D varies with input voltage and load, for conservative design in the above equation you may put D=D_{MAX}, where D_{MAX} - maximum duty cycle of the PWM (D_{MAX} <1).

Input power **Pin** is a function of output power **Pout** and efficiency. For example, if expected efficiency from the transformer input to the converter output is 95%, put Pin≈Pout/0.95.
Fill factor **K** depends on the coil construction. Typically, K=0.25...0.6 for wire-wound transformers and K=0.1...0.2 for planar cores.
Typical choice for current density is J=400-500 A/sq.cm. This value may need to be revised after measuring actual hot spot temperature rise in the coils. There is some degree of flexibility in choosing K and J.

The main **transformer design's constraint**: peak magnetic flux density B_{PK} should not reach the core material's saturation flux B_{SAT} at highest operation temperature. For bi-directional excitation (like in a full bridge) $B_{PK}=\Delta B/2$, for unidirectional excitation (like in a forward converter) $B_{PK}=\Delta B+B_r$, where B_r - core remanence. For customary 70% derating choose ΔB such that $B_{PK} \leq 0.7 \cdot B_{SAT}$. Note that in high frequencies, core loss rather than saturation can become the main limiting factor.

To reduce B_r and thus to increase available ΔB in single-ended topologies a small air gap can be introduced. Resulting B_r (in gauss) with a gap is:

$$Br = Hc \frac{\mu_r \cdot lm/lg}{\mu_r + lm/lg}$$

where Hc - coercive force (oersted), μ_r – relative permeability of ungapped core, lm - magnetic core path length (cm), lg – net length of the gap (cm).

For selected core based on required **Ap** and for given ΔB the primary turns:

$$N1 \geq \frac{V_{IN} \cdot Dmax}{F \cdot \Delta B \cdot 10^{-8} \cdot Ac}$$

Note. Input voltage normally varies over certain range from V_{IN_MIN} to V_{IN_MAX}, and the duty cycle varies inversely with the V_{IN}. With properly selected turns ratio, in steady state operation D_{MAX} would be reached only at low line V_{IN_MIN}. Therefore, for an aggressive design where minimum size is the main object, we can put V_{IN_MIN} instead of V_{IN} in the numerator of the above equation. If the converter has to meet certain holdup time, V_{IN_MIN} is the voltage to which input bulk capacitor discharges during such time. Note that during possible input/output transients the duty cycle may reach maximum even at high line, which will result in the flux exceeding the selected B_{PK}. Therefore, for a conservative design you may put V_{IN_MAX} in numerator of above equation.

Secondary turns are calculated based on the required turns ratio N1/N2 for a given converter topology (see Chapter 1).

4.2 CORE LOSSES

Core losses are proportional to area of the hysteresis loop, frequency and core volume. There is no general expression for analytical calculation of the core losses. There is an empirical Steinmetz approximation:

$$Pc = Kc \cdot F^{\alpha} \cdot B_{PK}^{\beta}$$

where Pc – specific core loss per unit volume; Kc, α and β – coefficients specific for given material.
In reality, these coefficients vary with frequency and temperature. For example, for Magnetics "R" material there is a curve fittin g approximation proposed in [8]:

$$P_C = (28.32 - 3.626 \cdot \ln F_{kHz}) \cdot F_{kHz}^{1.729} \cdot \Delta B^{(2.8332 - 0.00076 \cdot F_{kHz})}$$

where P_C – specific core losses at 100 °C in mW/cm^3, F_{kHz} - switching frequency in kHz, ΔB - flux swing in Tesla.

In practice, core losses are usually determined from the curves provided by magnetics manufacturers. These curves provide specific core loss per unit volume vs. peak flux density B_{PK}. The curves are normally given for bi-directional excitation with sinusoidal voltages. For such excitation B_{PK} is half the total flux swing: $B_{PK} = \Delta B/2$. In single-ended topologies we likewise have to use $\Delta B/2$ for B_{PK}. For example, for a single-ended excitation with passive reset, flux swing $\Delta B = B_{PK} - Br$. To apply the core loss curves for such topologies we use $(B_{PK} - Br)/2$.
Note that since in most topologies the voltage applied to power transformer is a square wave resulting in a triangular flux excitation, actual core losses slightly differ from those obtained from the curves.

4.3 COPPER LOSSES

Power loss in a wire generally consists of DC and AC losses. In a single-layer coil with sinewave AC component of the current, total losses are given by:

$$P_W = I_{RMS}^2 \cdot R_{DC} + I_{RMS\_AC}^2 \cdot R_{AC},$$

where I_{RMS} – total RMS current, I_{RMS_AC} - RMS value of AC component of the current, R_{DC} and R_{AC} - resistances of wire for DC and AC currents. The second term in P_W represents extra loss due to skin and proximity effects.

Coil's DC resistance:

$$R_{DC} \approx R_{DC/CM} \cdot MLT \cdot N$$

where $R_{DC/CM}$ - resistance in ohms per 1 cm length; MLT – mean length per turn of the bobbin (cm); N - turns.

$R_{DC/CM}$ can be obtained from the wire's datasheet, or can be calculated as the following:

$$R_{DC/CM} = \frac{\rho_{20}}{A_W} \cdot [1 + 3.9 \cdot 10^{-3} \cdot (T - 20)]$$

where $\rho_{20} = 1.724 \cdot 10^{-6}$ Ω·cm - copper resistivity at 20 °C; T - temperature (°C); A_W - cross-sectional area in cq.cm. For round wire $A_W = \pi \cdot D_W^2 / 4$, where D_W - bare wire diameter. Since initially you don't know the wire temperature rise, start with T=80-120 °C.

For standard American wire gauge AWG number:

$$D_W = 2.54 \cdot 10^{-\left(\frac{AWG+10}{20}\right)}$$

$$A_W \approx 5.067 \cdot 10^{-\left(\frac{AWG+10}{10}\right)}$$

SKIN EFFECT

Current at high frequencies moves to the surface ("skin") of the conductor. Penetration (skin) depth defined as the distance from the conductor surface to where the current density drops to $1/e$ of the surface current density.

For copper:

$$\delta_{SKIN} \approx (7.2...7.6)/\sqrt{F},$$

where δ_{SKIN} in cm, F – frequency in Hz.

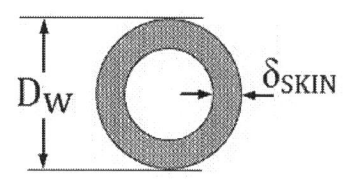

Effective cross-sectional area of wire that conducts current is:
$$A_{SKIN} \approx \pi \cdot \delta_{SKIN} \cdot (D_W - \delta_{SKIN})$$

Effective wire resistance R_{SKIN} due to skin effect is greater than the DC resistance R_{DC} by the factor:
$$\frac{R_{SKIN}}{R_{DC}} \approx \frac{\pi \cdot D_W^2 / 4}{A_{SKIN}} = \frac{D_W^2}{4 \cdot \delta_{SKIN} \cdot (D_W - \delta_{SKIN})}$$

PROXIMITY EFFECT

When conductors in close proximity carry currents in the opposite directions, the currents move towards the inside surfaces, which increases effective AC resistance. The ratio between AC and DC resistances in the layer "m" for sinusoidal currents in given by Dowell equation [9]:

$$\frac{R_{AC}}{R_{DC}} = X \cdot \frac{e^{2X} - e^{-2X} + 2\sin(2X)}{e^{2X} + e^{-2X} - 2\cos(2X)} + \frac{m^2 - 1}{3} \cdot 2X \cdot \frac{e^X - e^{-X} - 2\sin(X)}{e^X + e^{-X} + 2\cos(X)}$$

where $X = h/\delta_{SKIN}$ – ratio between height of the copper foil layer and the skin depth. For round wire use: $h = 0.8862 \cdot D_W$.

Classical Dowell curves provide R_{AC} normalized to R_{DC} as a function of X. Since R_{DC} itself varies with h, these curves do not show how actual value of R_{AC} varies with the height of the copper layers. It is more useful to normalize R_{AC} to $R_{DC\delta}$ - DC resistance of a foil of thickness $h = \delta_{SKIN}$. Since DC resistance of foil is inverse proportional to its height, actual value of R_{AC} varies as following:

$$R_{AC} = \frac{R_{AC}}{R_{DC}} \cdot R_{DC} = \frac{R_{AC}}{R_{DC}} \cdot R_{DC\delta} \cdot \frac{\delta_{SKIN}}{h}$$

where R_{AC}/R_{DC} is given by Dowell equation.
Here are R_{AC} curves normalized to $R_{DC\delta}$:

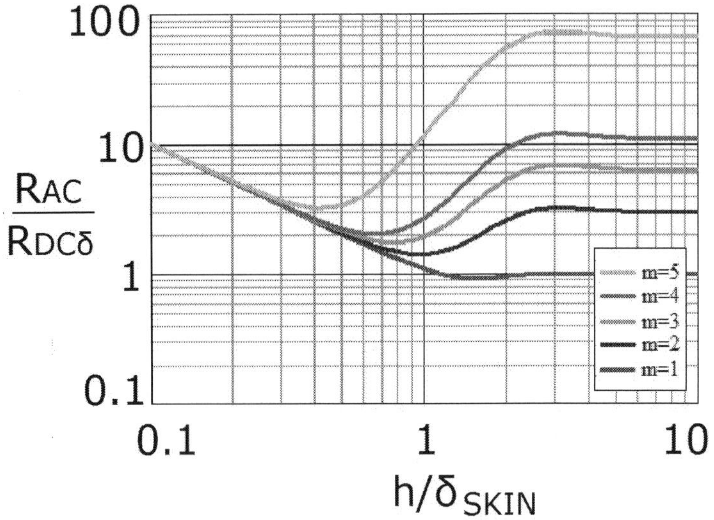

For non-sinusoidal currents R_{AC} is provided in [10].

TEMPERATURE RISE

Empirical formula for hot-spot temperature rise in °C vs. total power losses P(watt) in transformer [11]:

$$\Delta T \approx 50 \cdot \frac{P}{\sqrt{Ve}}$$

where Ve - core volume in cubic centimeters.

4.4 POWER INDUCTOR DESIGN

Peak magnetic field strength (gauss):

$$H_{PK} = 0.4 \cdot \pi \cdot N \cdot I_{PK} / lm$$

where I_{PK} – peak current (DC current plus $\Delta I_L/2$); N – turns, lm – effective magnetic path length (cm).

Inductance of a coil (henry):

$$L = 0.4\pi \cdot \mu_{eff} \cdot N^2 \cdot \frac{Ac}{lm} \cdot 10^{-8}$$

where Ac - effective core's cross-sectional area (sq.cm); μ_{eff} – core's effective relative permeability.

The magnetic flux (gauss) in general: $B = \mu_{eff} \cdot H$.

Substituting H_{PK} and eliminating lm yields the required $N \cdot Ac$ product of a "current-driven" coil:

$$N \cdot Ac = \frac{L \cdot I_{PK} \cdot 10^8}{B_{PK}}$$

where B_{PK} – peak flux.

We normally choose $B_{PK} \leq 0.7 \cdot B_{SAT}$ for 70% derating, where B_{SAT} - saturation flux at highest operation temperature.

The cross-sectional area Ac is determined by selected core size, which is a function of desired current density in the coil:

$$J = N \cdot I_{RMS} / (Aw \cdot K)$$

where J – current density (amps/sq.cm), Aw - window area (sq.cm), K – winding fill factor (typical $K = 0.25...0.6$), I_{RMS} - RMS value of the inductor current.

Eliminating N from the above equations yields required area product:

$$Ap = Ac \cdot Aw = \frac{L \cdot I_{PK} \cdot I_{RMS}}{K \cdot J \cdot B_{PK} \cdot 10^{-8}}$$

Magnetic areas Ac are provided in the datasheets, Aw values can be found in the matching bobbins datasheets. The core size is selected based on the required Ap.

There is some degree of flexibility in choosing K and J.
Typical selection of current density is J=400-500 A/sq.cm. It may need to be revised after measuring actual hot spot temperature rise in the coil.
For selected core the required turns:

$$N = \frac{L \cdot I_{PK} \cdot 10^8}{B_{PK} \cdot A_c}$$

To prevent the magnetic material saturation at I_{PK}, we usually introduce an air gap.
The net air gap (in cm) required to limit B_{PK} at I_{PK}:

$$lg \approx \frac{0.4 \cdot \pi \cdot N \cdot I_{PK}}{B_{PK}} - \frac{lm}{\mu_r}$$

where μ_r – relative permeability of ungapped core.
Effective relative permeability for ferrite cores with a discrete air gap:

$$\mu_{eff} = \mu_r \frac{lg + lm}{\mu_r \cdot lg + lm}$$

Inductance of gapped core:

$$L = 0.4\pi \cdot N^2 \cdot \frac{A_c}{lg + lm/\mu_r} \cdot 10^{-8} \approx 0.4\pi \cdot N^2 \cdot \frac{A_c}{lg} \cdot 10^{-8}$$

For powder metal materials with a distributed gap and soft saturation curve, the calculation process may take several iterations. In short, you can first pick a powder core based on desired $L \cdot I_{PK}^2$ by using manufacturer's charts. Then determine the initial number of the turns:

$$N = \sqrt{\frac{L}{A_L \cdot 10^{-9}}}$$

where **AL** - specific inductance in **mH/1000** turns (which is **nH/turn**) from the core's data sheet.

For selected turns find peak bias $H = 0.4 \times \pi \times N \times I_{PK} / l_m$ (Oersted) and determine the permeability roll-off in percentage of initial permeability at such bias. Then if needed, increase the turns for the desired L.

5 MISCELLANEOUS POWER ELECTRONICS REFERENCE

5.1 ESTIMATION OF MOSFET LOSSES

CONDUCTION LOSSES

If gate voltage $V_g > V_{th}$ (where V_{th} – gate threshold), and external load is such that the drain to source voltage $V_{ds} < (V_g - V_{th})$, the drain-to-source channel acts like a nearly constant resistance R_{dson}. In such mode the drain current I_D is determined by external circuit's impedance and the voltage source. Power dissipation then is $P = I_{D\_RMS}^2 \cdot R_{dson}$. The R_{dson} has a positive temperature coefficient 0.5-1.0% per $^\circ C$ at the junction, and may double from 25 $^\circ C$ to 125 $^\circ C$. It slightly drops as gate voltage increases above V_{th}.

SWITCHING LOSSES

Switching losses are functions of three internal capacitors (gate to source Cgs, drain to gate Cdg, and drain to source Cds), internal diode charge, and gate drive current. MOSFET data sheet usually provide equivalent capacitances seen from each terminal specified at a certain voltage V_{DS_SPEC}:

- Input capacitance $C_{iss} = C_{gs} + C_{dg}$ (drain shorted to source);
- Output capacitance $C_{oss} = C_{ds} + C_{dg}$ (gate shorted to source);
- Reverse transfer or Miller capacitance $C_{rss} = C_{dg}$.

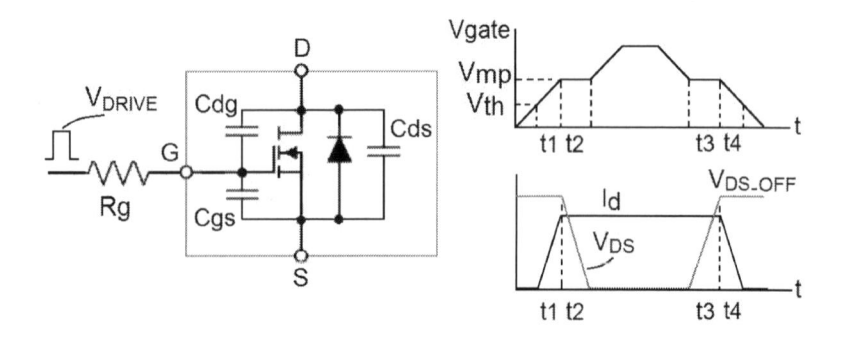

C_{oss} and C_{rss} are functions of the actual drain to source voltage of a MOSFET. Average values of these capacitances [12]:

$$C_{RSS\_AV} = 2C_{RSS} \cdot \sqrt{\frac{V_{DS\_SPEC}}{V_{DS\_OFF}}}$$

$$C_{OSS\_AV} = 2C_{OSS} \cdot \sqrt{\frac{V_{DS\_SPEC}}{V_{DS\_OFF}}}$$

where V_{DS_OFF} – MOSFET voltage in off-state.

In general, switching losses depends on the topology. Net switching losses with inductive load:

$$P_{sw} \approx \frac{I_d \cdot V_{DS\_OFF} \cdot (t1 + t2 + t3 + t4) + C_{OSS\_AV} \cdot V_{DS\_OFF}^2 \cdot F}{2}$$

where I_d – drain current, F – switching frequency; $t1$ – turn-on delay (rise time of gate voltage from gate threshold V_{th} to Miller plateau V_{mp}); $t2$ – the drain voltage fall time (Miller plateau duration); $t3$ – voltage rise time during turn off; $t4$ – current fall time [13]:

$$t1 = R_g \cdot C_{iss} \cdot \ln\left(\frac{V_{DRIVE} - V_{th}}{V_{DRIVE} - V_{mp}}\right)$$

where R_g – net gate drive resistance; V_{DRIVE} – gate drive voltage;

$$t2 = \frac{V_{DS\_OFF} \cdot R_g \cdot C_{rss}}{V_{DRIVE} - V_{mp}}$$

$$t3 = R_g \cdot C_{rss} \cdot \frac{V_{DS\_OFF}}{V_{mp}}$$

$$t4 = R_g \cdot C_{iss} \cdot \frac{V_{mp}}{V_{th}}$$

5.2 CALCULATION OF OUTPUT VOLTAGE RIPPLE

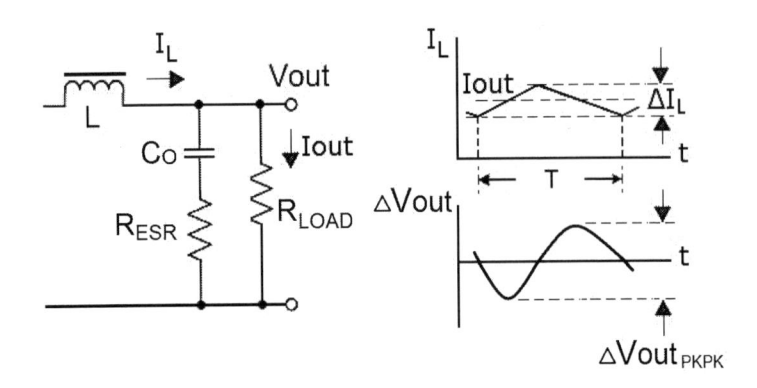

Output voltage ripple ΔVout consists of the ripple due to ESR and the ripple due to the capacitor charge/discharge. The peaks of these two components occur at different times, and the resulting ripple is a function of duty cycle. Neglecting DC resistance of the inductor and ESL of the capacitor, peak to peak output ripple for buck-derived topologies in CCM can be approximated as following [14]:

$$\Delta V_{out\,PKPK} \approx \frac{\Delta I_L}{8 C_o \cdot F} + \Delta I_L \cdot R_{ESR}$$

where $F=1/T$ – switching frequency (hertz), C_o – capacitance (farad), R_{ESR} – equivalent series resistance of C_o, ΔI_L – peak to peak inductor ripple current (see Chapter 1 for ΔI_L in various topologies).

Output ripple are usually selected <1% of output voltage. It is customary to design the filter such that each component of the ripple contributes to half of total ripple.

$$C_o \geq \frac{1}{\Delta V_{out\,PKPK}} \cdot \frac{\Delta I_L}{4 \cdot F}$$

$$ESR \leq \Delta V_{out\,PKPK} / 2\Delta I_L$$

5.3 CAPACITANCE CALCULATION FOR LOAD TRANSIENT RESPONSE

Since inductor can't change its current instantaneously, during rapid load change transient current flows through output capacitors before the control loop responds. This results in transient output voltage deviation (overshoot/undershoot).

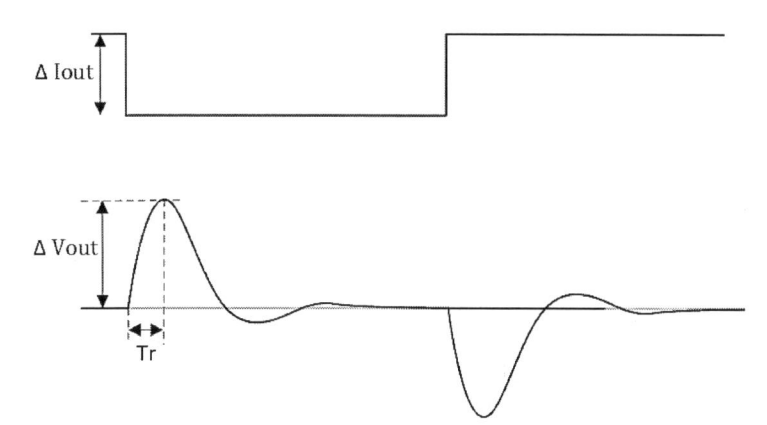

At small load steps $\Delta Iout$ when all loop components remain within their linear region (small-signal operation), output excursion is:

$$\Delta Vout \approx \Delta Iout \cdot |Z_{CL}(Fc)|$$

where $Z_{CL}(Fc)$ – output impedance of a closed-loop circuit at crossover frequency Fc.

Negative feedback reduces effective output impedance, which can reduce output voltage transient deviation, but beyond crossover frequency the feedback does not help. In general, output impedance of closed-loop circuit with negative feedback is:

$$Z_{CL}(s) = Z_{OL}(s)/(1+T_{OL}(s))$$

where Z_{OL} - open-loop output impedance, $T_{OL}(s)$ - open loop transfer function:

$$T_{OL}(s) = |T_{OL}(s)| \cdot e^{j\varphi}$$

where $\varphi = \arg\{T_{OL}(j\omega)\}$.

Assuming small ESR of output capacitors: $Z_{OL}(s) \approx 1/s \cdot C_o$.
At crossover frequency $|T_{OL}(s)|=1$ and closed loop impedance is:

$$|Z_{CL}| \approx 1/2\pi F_c \cdot C_o$$

Required output capacitance to meet step load voltage transient requirements for selected crossover frequency:

$$C_o > \frac{\Delta I}{2\pi F_c \cdot \Delta V_{out}}$$

Note that if the crossover frequency F_c is higher than $1/(2\pi \cdot C_o \cdot ESR)$, peak ΔV_{out} occurs at the first moment and is determined by equivalent ESR of output capacitors: $\Delta V_{out} = \Delta I_{OUT} \cdot ESR$ [5].

RESPONSE TO FINITE SLEW RATE OF THE LOAD

The response time to load step $T_r \approx 1/(4 \cdot F_c)$. With an ideal step load, the charge removed from or delivered to C_o during this time:

$$\Delta Q \approx 0.5 \cdot \Delta I_{out} \cdot T_r$$

In practice, load transitions occur over some non-zero time. With a finite slew rate di/dt assuming load change is much faster than the response time:

$$\Delta Q \approx 0.5 \cdot \Delta I_{out} \cdot (T_r - \frac{\Delta I_{out}}{di/dt})$$

Output voltage excursion in general is: $\Delta V_{out} \approx \Delta Q/C_o$.
Then the required output capacitance:

$$C_o > \Delta Q/\Delta V_{out} \approx 0.5 \cdot \frac{\Delta I_{out}}{\Delta V_{out}} \cdot (\frac{1}{4 \cdot F_c} - \frac{\Delta I_{out}}{di/dt})$$

For the selection of crossover frequency see par.3.4. As a starter, you may select F_c as 1/10 of switching frequency.

5.4 PCB TRACE PROPERTIES

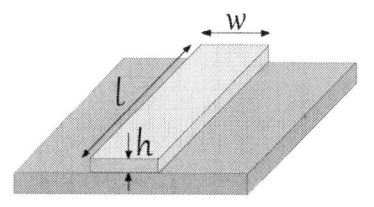

PCB TRACE RESISTANCE

$$R = 1.7 \cdot 10^{-6} \cdot \frac{l}{w \cdot h} [1 + 0.0039 \cdot (T - 25)]$$

where R – net resistance in ohms, l – trace length (cm), w – width (cm), h – copper thickness (cm), T – temperature (Celsius).
If copper is specified in ounces as it's common to U.S., you can find trace thickness in centimeters as: h≈(0.0033-0.0035)·W_{OZ}, where W_{OZ} – copper weight in ounces per square foot (ounces/sq.ft). For example, 1 ounce copper has average thickness 0.0034 cm (1.3 mils).

PCB TRACE SELF INDUCTANCE

$$L \approx 0.002 \cdot l \cdot (\ln \frac{2 \cdot l}{w + t} + 0.5)$$

where L – self-inductance in µH; all dimension are in cm.

REQUIRED TRACE WIDTH vs CURRENT

An approximation obtained from IPC-2152 curves of current vs. cross-sectional area [15]:

$$w = \frac{0.769}{W_{OZ}} \cdot (117.555 \cdot \Delta T^{-0.913} + 1.15) \cdot I^{1.159 + 0.84 \cdot \Delta T^{-0.108}}$$

where w – required trace width in mils for temperature rise ΔT (°C) at rms current I (amps).

TEMPERATURE RISE

Empirical temperature rise at PCB surface for heat dissipation by convection with no forced air flow [16]: $\Delta T(°C) = 133 \cdot P/A$,
where P – net power dissipated on the surface of the board (watts), A – board surface area (sq.in).

Typical **dielectric withstand voltage** between uncovered PC traces or components on a clean board: 40 V/mil ($\approx$1600 V/mm).

6 REFERENCES

1. Hong Huang, "Designing an LLC Resonant Half-Bridge Power Converter", 2010 Texas Instruments Power Supply Design Seminar, SLUP263.
2. R. W. Erickson, Bode Diagrams of Transfer Functions and Impedances, ECEN 2260, 1997.
3. C. Basso, "Eliminate the guesswork in crossover frequency selection", Power Electronics and Technology, August 2008, pp. 24-29.
4. R. W. Erickson and D. Maksimovic, Fundamentals of Power Electronics, 2nd ed., Norwell, MA: Kluwer, 2000.
5. K. Yao, Y. Ren and F. C. Lee, "Critical bandwidth for the load transient response of voltage regulator modules," IEEE Transactions on Power Electronics, Vol. 19, No. 6, November 2004, pp 1454- 1461.
6. R.B. Ridley, "A new, continuous-time model for current-mode control," IEEE Trans. on Power Electronics, vol. 6, no. 2, April 1991, pp. 271-280.
7. C. Basso, "Designing Control Loops for Linear and Switching Power. Supplies: A Tutorial Guide", Artech House, 2012.
8. Ray Ridley and Art Nace, "Modeling ferrite core losses," Switching Power Magazine, vol. 3, no. 1, pp. 6–13, 2002.
9. Dowell, P.L., Effects of eddy currents in transformer windings, IEE Proceedings, vol. 113, No. 8, August, 1966, pp. 1387–1394.
10. B. Carsten, "High Frequency Conductor losses in Switchmode Magnetics," HFPC Proceedings, 1986.
11. Mulder S.A., "Application note on the design of low-profile high frequency transformers", Philips, 1990.

12. "Estimating MOSFET Parameters From the Datasheet", Appendix A. Texas Instrument Seminar Topics SLUP170, 2002.
13. Brown, "Power MOSFET Basics: Understanding Gate Charge and Using It To Assess Switching Performance", Vishay Siliconix AN-608, 2004.
14. Singh, S.P. "Output ripple voltage for buck switching regulator". Application Report SLVA630A, Texas Instruments, Inc, 2014.
15. Rozenblat L., "PCB Trace Width Calculator and Equations", https://www.smps.us/pcb-calculator.html, 2012.
16. Mauney, Charles, "Thermal Considerations for Surface Mount Layouts", Texas Instruments.

7 APPENDIX

7.1 MAGNETIC UNIT CONVERSION

QUANTITY	SI UNIT	CGS UNIT	RELATION
Magnetic induction B	Tesla (T)	Gauss (G)	$1\,T = 10^4\,G$
Magnetic field strength H	Ampere/meter (A/m)	Oersted (Oe)	$1\,A/m = 4\pi \cdot 10^{-3}\,Oe \approx 1/80\,Oe$
Magnetic flux Φ	Weber (Wb)	Maxwell (M)	$1\,Wb = 10^8\,M$
Magnetization M	Ampere/meter (A/m)	emu/cm^3	$1\,A/m = 10^{-3}\,emu/cm^3$

MAGNETICS EQUATIONS IN DIFFERENT UNIT SYSTEMS

Magnetic induction
SI: $B = \mu_0 \cdot (H+M) = \mu_{eff} \mu_0 \cdot H$,
where $\mu_0 = 4\pi \cdot 10^{-7}$ - magnetic permeability of vacuum; μ_{eff} – effective relative core permeability; M - magnetization (magnetic moment per unit volume);
CGS: $B = H + 4\pi \cdot M = \mu_{eff} \cdot H$.

Magnetic field strength
SI: $H = N \cdot I / l_c$, where l_c – magnetic path, meters;
CGS: $H = 0.4\pi \cdot N \cdot I / l_c$, where l_c - magnetic path, cm

8 ABOUT THE AUTHOR

Lazar Rozenblat is a retired electrical engineer with over 30 years of experience in the practical power electronics design. Particularly, he was a manager of R&D group at Todd Products, principal engineer at Transistor Devices (TDI) R&D center, and principal applications engineer at Microchip. Lazar holds an US Patent in Power-over-Ethernet field and authored a number of technical papers on power supply design. He and his wife have two daughters and so far five grandchildren.

Made in the USA
Middletown, DE
27 September 2023